Enhancement by Enlargement

The Proliferation Security Initiative

Charles Wolf, Jr., Brian G. Chow,
Gregory S. Jones

Prepared for the Office of the Secretary of Defense

Approved for public release; distribution unlimited

NATIONAL DEFENSE RESEARCH INSTITUTE

The research described in this report was prepared for the Office of the Secretary of Defense (OSD). The research was conducted in the RAND National Defense Research Institute, a federally funded research and development center sponsored by the OSD, the Joint Staff, the Unified Combatant Commands, the Department of the Navy, the Marine Corps, the defense agencies, and the defense Intelligence Community under Contract W74V8H-06-C-0002.

Library of Congress Cataloging-in-Publication Data is available for this publication.

ISBN: 978-0-8330-4579-9

Published 2008 by the RAND Corporation
1776 Main Street, P.O. Box 2138, Santa Monica, CA 90407-2138
1200 South Hayes Street, Arlington, VA 22202-5050
4570 Fifth Avenue, Suite 600, Pittsburgh, PA 15213-2665
RAND URL: http://www.rand.org/
To order RAND documents or to obtain additional information, contact
Distribution Services: Telephone: (310) 451-7002;
Fax: (310) 451-6915; Email: order@rand.org

Preface

The Proliferation Security Initiative (PSI) is of growing importance in a world where access to technology, components, and weapons of mass destruction is also growing. The purpose of the research behind this monograph is to enhance PSI's effectiveness by acknowledging, understanding, and, if possible, overcoming the reluctance of several key countries to endorse and participate in the initiative's security-enhancing efforts.

This research was sponsored by the Office of the Secretary of Defense and conducted within the International Security and Defense Policy Center of the RAND Corporation's National Defense Research Institute, a federally funded research and development center sponsored by the Office of the Secretary of Defense, the Joint Staff, the Unified Combatant Commands, the Department of the Navy, the Marine Corps, the defense agencies, and the defense intelligence community.

For more information on RAND's International Security and Defense Policy Center, contact the Director, James Dobbins. He can be reached by email at James_Dobbins@rand.org; by phone at 703-413-1100, extension 5134; or by mail at the RAND Corporation, 1200 S. Hayes Street, Arlington, VA 22202. More information about RAND is available at www.rand.org.

Contents

Summary

The Proliferation Security Initiative (PSI), begun in 2003, was conceived as an *activity* rather than an organization, the intention being to focus on collective action while avoiding the bureaucratic impediments that organizations often entail. PSI's purpose is to prevent or at least inhibit the spread of weapons of mass destruction (WMD), their delivery systems, and related materials to or from states or non-state actors whose possession of such items would be a serious threat to global or regional security. An Operational Experts Group (OEG) of 20 countries leads the initiative's operations, planning and implementing the exercises and other multilateral efforts designed to further PSI's purpose. Ninety-one countries, including the OEG members, make up this group of widely multilateral participants, all of which have endorsed PSI's purpose and principles.

This RAND project for the Office of the Secretary of Defense's Policy Office had two objectives. The first was to assess the advantages and disadvantages, or benefits and costs, that, when balanced against each other, have induced five key countries not to affiliate with PSI publicly, and to ascertain whether (and, if so, how) this balance might be altered to enhance the prospects for their affiliation in the near future. Implicit in this objective is the premise that PSI's effectiveness will be enhanced by enlarging the number of participants.

The project's second objective was to develop a syllabus of training materials, partly by drawing on work done in connection with the first objective. The syllabus is intended to help U.S. Geographic Combat Commands mitigate problems arising from normal staff turnover and insufficient institutional memory, and thereby to improve the commands' ability to provide operational support for the numerous multilateral exercises constituting the core of PSI's peacetime activities.

We address the first objective in this report; the second will be addressed in a separate document that also provides additional details about PSI and about relevant treaties, agreements, and programs discussed in this report.

The five countries of interest—Indonesia, Malaysia, Pakistan, India, and China—share an implicit calculus that the costs (disadvantages) associated with PSI affiliation exceed, or at least equal, the benefits (advantages). We identify specific issues within these countries' assessments for which the benefits ascribed to PSI may have been underestimated and/or the costs ascribed to PSI may have been overestimated. We then suggest how these under- and overestimates might be changed in ways that would lead these countries to reconsider their decision not to affiliate with PSI. Of course, if these countries see their estimates as correct and not subject to reconsideration, it follows that their nonaffiliated PSI status will remain unchanged.

The Five Countries

We begin by dividing the five countries into three groups: Indonesia and Malaysia; Pakistan and India; and China. These groupings reflect the conjecture that the probability of one member in a two-member group (the group of one does not play in this conjecture) changing its stance of nonaffiliation with PSI is likely to be affected by whether the other member of that group alters its stance. However, these groupings preclude neither the possibility of interactions between countries in the different groups, nor the possibility of significant interactions with countries other than these five. Indeed, interactions between Saudi Arabia, which endorsed PSI in May 2008, and India and Pakistan remain relevant to India's and Pakistan's assessments of whether to endorse PSI.

In analyzing the five countries' decisions, we have attempted to adopt their separate perspectives and sensitivities in order to better understand why or how each may have overestimated the disadvantages and/or underestimated the advantages of PSI affiliation. For example, to the extent that China views PSI as a U.S.-dominated activity and continues seeking to strengthen the Shanghai Cooperation Organization as a counterweight to the United States in Asia, China may conclude that nonaffiliation is the preferable stance.

In considering the first group, Indonesia and Malaysia, we begin by describing their interdependencies and shared interests. We then turn to three salient issues and concerns they both have that have so far led them to refrain from formal PSI affiliation: sovereignty, law of the sea, and independent foreign policy.

The two members of the second group, Pakistan and India, are current nuclear powers that, in certain circumstances, might have reasons and resources that would dispose them to assist Saudi Arabia if it sought to acquire a nuclear capability of its own (perhaps in response to such an acquisition by Iran). One of the several intricate interactions among the three is that the nexus between the possible interest in future acquisition by the Saudis and the possible sources of future supply represented by Pakistan or India might influence Pakistan and India to avoid or at least defer PSI affiliation. However, in the case of India, internal political circumstances are currently much more formidable obstacles to joining PSI.

We chose to treat China separately for several reasons. In addition to being a nuclear weapon state, it is the second or third largest economy in the world, the fourth or fifth largest global trading country, and the third or fourth largest global weapons exporter. Moreover, it has a mixture of political, economic, and security interests and transactions with North Korea and Iran, the two major current and prospective sources of "proliferation concern" in the world. The mixture and complexity of interests at stake for China include a prevalent belief among its leadership that blandness and "carrots" rather than coercion and "sticks" enhance its ability to influence North Korea, and that affiliation with PSI would, by appearing threatening to North Korea, compromise this ability.

China's inclination toward Iran is similar—it seeks to temporize rather than pressure. China's estimate of the consequences of PSI affiliation may also be influenced by reluctance to jeopardize its substantial and growing trade and investment transactions with Iran.

The Five Principles

After a broad assessment of the general and specific benefits that countries typically associate with PSI affiliation, we address the key policy question of this research: What measures, policies, and approaches can the United States and other PSI participants invoke that are likely to induce each of the five countries to lower its estimates of PSI-affiliation costs (disadvantages) and/or raise its estimates of PSI-affiliation benefits (advantages) such that it arrives at a positive (rather than negative or neutral) bottom-line estimate?

To assist us in answering this question, we set out five general principles to use as guides in seeking remedial policies conducive to PSI affiliation by the five countries. Each principle applies to at least one of the five countries; several apply jointly to more than one country. These principles are as follow:

1. Exercising U.S. leadership by ceding it to other PSI participants
2. Interpreting and applying "innocent passage" consistent with each state's own national legal authorizations and its obligations under international law
3. Affirming the validity of "territorial waters" and emphasizing the locus of responsibility in the littoral countries
4. Presenting PSI affiliation as incremental to agreements and/or commitments already arrived at
5. Conferring membership in the Operational Experts Group (OEG) ab initio.

Note that our five principles do not include carrots and sticks related to issues outside PSI, such as peaceful nuclear assistance for nuclear power plants. Instead, they focus on assuring the five countries that PSI participation will not interfere with their existing international obligations and rights, which should enable them to reassess the costs and benefits of PSI affiliation. Our objective is for the five to affiliate with PSI because they consider the benefits of doing so to outweigh the costs—not because they want to use the act of affiliation as a bargaining chip for obtaining benefits or avoiding penalties on issues unrelated to PSI or nonproliferation. Affiliation for its own sake will make them more-active participants and, in the long run, will meet the nonproliferation objective far better than affiliation for extraneous reasons.

Applying the Five Principles

Our next step was to apply the principles to each of the countries, in the process considering how application might alter the calculus of costs and benefits of PSI affiliation. The aim and the result of this exercise were the same: to suggest the manner in which each country should be approached and the points that should be highlighted in inviting each one to join PSI.

Indonesia and Malaysia

It would be prudent for the PSI invitations extended to Indonesia and Malaysia to come from Singapore, Japan, France, Australia, and one or two states in the Gulf Cooperation Council (GCC) (principle 1). Acting on behalf of the full PSI constituency, these countries would explicate the subjectivity of determining what may or may not be innocent passage (principle 2) and the unambiguous protection of territorial waters by the littoral states (principle 3). New Zealand might usefully be included among the several countries extending the invitation, partly because of its geographic proximity and partly because it has effectively articulated the broad scope of benefits from PSI affiliation.

Given these two countries' viewpoints on the United States, it may be advisable to pursue this approach initially with Malaysia and then with Indonesia. Moreover, approaching Malaysia first would benefit from the fact that Malaysia has already been an observer in three PSI exercises and has joined the Container Security Initiative (CSI). This is an application of principle 4 for building on the five countries' relevant prior activities, including their recent efforts to enact domestic laws and to join international agreements for nonproliferation. Although Indonesia has not observed any PSI exercises or joined CSI, it has made recent efforts along similar lines, both domestically and internationally, in support of nonproliferation. Moreover, because of Indonesia's stature and strategic location, its invitation should include an offer of immediate membership in the OEG.

India and Pakistan

It may be advisable to have France, the United Kingdom, and Russia—the three nuclear-state PSI participants other than the United States—and perhaps Japan, convey invitations to India and Pakistan. Having the United States forgo this role of formal protagonist (principle 1) may help allay India's sensitivity by emphasizing the multilateral character of PSI's activities and modulating the U.S. role in them.

The protagonists should assure India and Pakistan (and the other three countries, as well) that PSI will not compromise their right of innocent passage (principle 2). The incremental character of PSI affiliation should be emphasized in light of these countries' prior efforts to enact domestic laws and to join international agreements for nonproliferation (principle 4). For example, Pakistan has already participated as an observer in three PSI exercises, India in two. Finally, it would be appropriate and perhaps more effective if the invitation to both India and Pakistan were accompanied by an option for immediate participation in the PSI's OEG (principle 5).

China

China's affiliation with PSI should be sought by several principal PSI members, including but not confined to the United States. The invitation's effectiveness would be enhanced if, for example, it were extended jointly by France, the United Kingdom, Germany, and the United States, with the first three playing the lead role (principle 1). France and perhaps Russia, as another PSI member, might authoritatively convey the consistency between PSI affiliation, on one hand, and the appropriate and reasonable qualifications within PSI's interdiction principles that can be invoked to protect the right of genuinely innocent passage, on the other (principle 2).

With China, the United States may be in the best position to explicate the incremental and complementary nature of PSI affiliation (principle 4). In the last two decades, China has taken part in many international nonproliferation treaties and agreements. Also, China has already placed three of its principal ports (Hong Kong, Shanghai, and Shenzhen) under the purview of CSI. Consequently, PSI can be accurately portrayed as only a modest additional step that complements China's other nonproliferation efforts. The persuasiveness of China's invitation is likely to be enhanced by having all four of the inviting powers extend the option of immediate status in the OEG upon affiliation (principle 5).

Preliminary Ideas for Further Consideration

We also provide, for further consideration, some preliminary ideas on PSI's development and, more specifically, on the pros and cons of PSI affiliation:

- Discussing with the insurance industry whether and, if so, how premiums charged for insuring cargo (whether transported by surface, air, or sea) take into account any risk abatement related to affiliation with PSI of the transport vehicle's nation of origin.
- Considering ways to allay concerns about the right of innocent passage, especially the concern that an innocent ship might suffer delay because of interdiction.
- Clarifying possible misinterpretation about the relationship between the United Nations Convention on the Laws of the Sea (UNCLOS) and PSI with respect to the right of innocent passage, including appropriate rules of engagement that would reassure littoral states that their prerogatives in their own territorial seas would not be infringed upon by PSI interdiction principles.
- Considering whether to offer prospective PSI members technical assistance, inspection equipment, and other items that might help

improve their import/export control, inspection, and interdiction capabilities.
- Analyzing the status and trends of technology for sensing and detecting WMD that may enable better and quicker identification of WMD components, thereby enhancing the effectiveness of PSI.

Acknowledgments

We are pleased to acknowledge the helpful comments we received from RAND colleague David Mosher and from Henry Sokolski on an earlier draft of this study, and fully absolve them of any responsibility for the views expressed and conclusions reached in the final text.

Abbreviations

ASEAN	Association of Southeast Asian Nations
BCN	biological, chemical, or nuclear
CBP	U.S. Customs and Border Protection
CPI(M)	Communist Party of India (Marxist)
CSI	Container Security Initiative
DHS	Department of Homeland Security
EXBS	Export Control and Related Border Security Assistance Program
GCC	Gulf Cooperation Council
GDP	gross domestic product
IAEA	International Atomic Energy Agency
IMO	International Maritime Organization
NPT	Nuclear Nonproliferation Treaty (formally, the Nonproliferation of Nuclear Weapons Treaty)
OEG	Operational Experts Group
PLA	People's Liberation Army
PSI	Proliferation Security Initiative
SAREX	search-and-rescue exercises

SCO	Shanghai Cooperation Organization
SFI	Secure Freight Initiative
SUA	Suppression of Unlawful Acts (of Violence Against the Safety of Maritime Navigation)
UN	United Nations
UNCLOS	United Nations International Convention on the Law of the Sea
UNSCR	United Nations Security Council Resolution
UPA	United Progressive Alliance
WMD	weapons of mass destruction
WMD/missile items	weapons of mass destruction, their delivery systems, and related materials

CHAPTER ONE

Introduction

The Proliferation Security Initiative (PSI), currently in its fifth year, was conceived as an *activity* rather than an organization, signifying the intention to minimize bureaucratic and overhead burdens and underscoring the initiative's collective, multilateral character. PSI's purpose is to prevent or at least inhibit the spread of weapons of mass destruction (WMD), their delivery systems, and related materials[1] to and from states and non-state actors whose possession would be a serious threat to global or regional security. PSI's purpose thus involves provision of a "public good," whose "public" character partly derives from the activity's benefits accruing to all PSI affiliates regardless of whether or how they contribute to meeting its costs.[2]

PSI's functions are led by the Operational Experts Group (OEG), a group of military, law enforcement, intelligence, legal, and diplomatic experts from 20 of the countries participating in PSI. The OEG "meets regularly to develop operational concepts, organize the interdiction exercise program, share information about national legal authorities, and pursue cooperation with key industry sectors."[3] OEG meetings

[1] For convenience, we refer to all of these collectively as WMD/missile items in this report.

[2] We say "partly" here because the "public good" provided by PSI may benefit all transport, regardless of whether the shipper's flag is that of a PSI participant.

[3] U.S. Department of State, Bureau of International Security and Nonproliferation, *The Proliferation Security Initiative (PSI)*, Fact Sheet, Bureau of International Security and Nonproliferation, Washington, D.C., May 26, 2008. Details on PSI and its operations can be found in this fact sheet, which is replicated in full in Appendix A.

have been chaired by the countries hosting the PSI exercises. Membership in the OEG is voluntary, countries have been added to the group periodically, and decisions about projects and exercises are made cooperatively and collegially.

Under PSI, over 30 maritime, air, and ground interdiction exercises involving over 70 nations have been conducted. These exercises are hosted by individual PSI-participant countries throughout the world to improve coordination mechanisms in support of interdiction-related decisionmaking.

Although PSI was initiated by the United States, it has become impressively multilateral and now has 91 participants. Participation depends solely on general endorsement of the initiative's interdiction principles,[4] and each country is given wide latitude on its level of participation.

The two objectives of our study were

1. to assess the incentives and disincentives, or advantages and disadvantages, that on balance have induced five key countries not to affiliate with PSI, and to ascertain whether and, if so, how the balance, as assessed by the five, might be altered to enhance the prospects for their affiliation in the near future. Implicit in this objective is the premise that PSI's effectiveness will be enhanced by enlarging the number of countries that participate in its activities. This premise is not unique to PSI and, indeed, applies no less to other collective endeavors, such as those of the UN, NATO, and the Plaza Accords.
2. to develop a syllabus of training materials partly by drawing on the material in this report. The syllabus is intended to help U.S. Geographic Combat Commands mitigate problems arising from normal staff turnover and insufficient institutional

[4] U.S. Department of State, Bureau of International Security and Nonproliferation, *Proliferation Security Initiative: Statement of Interdiction Principles*, Fact Sheet, The White House, Office of the Press Secretary, September 4, 2003.(We present this statement of interdiction principles in full in Appendix B of this report.) Further analysis of the interdiction principles is contained in the syllabus we developed for a PSI training manual to meet the second objective of the overall study.

memory, thereby improving the commands' ability to provide operational support for the numerous multilateral exercises constituting the core of PSI's peacetime activities.

We address the first objective below; the second objective will be addressed in a separate report that cross-references relevant parts of this report.

CHAPTER TWO

Analysis of Costs and Benefits of PSI Affiliation for the Five Countries

Despite their extreme diversity, the five countries we evaluated—Indonesia, Malaysia, Pakistan, India, and China—have at least one obvious attribute in common: an implicit calculus that the costs (disadvantages) associated with PSI affiliation exceed or at least equal the benefits (advantages). The specific costs and benefits may differ for each of the five, but the bottom line that each arrives at is the same in that it is either negative or zero.

Our task, then, was first to identify specific items or issues in these countries' assessments whose benefits may have been underestimated and/or whose costs may have been overestimated, and then to suggest how these under- and overestimates might be changed so that the five countries would reconsider their decision not to affiliate with PSI.

Given these objectives, we found it useful to group the five countries as follows—Indonesia and Malaysia, Pakistan and India, and China by itself—based on a specific criterion: the existence of current and prospective interactions between countries that may affect the probability of one country reversing its decision not to affiliate with PSI or at least deciding to cooperate more actively with PSI. In other words, we posited that the probability of one member of a group changing its stance would be affected by whether the other member of that group altered its stance of nonaffiliation with PSI. This grouping does not preclude the possibility of interactions between members of different groups, nor does it exclude the possibility of significant interactions with countries other than these five.

Indonesia and Malaysia

Both Indonesia and Malaysia share major interests and characteristics. They are founding members of the Association of Southeast Asian Nations (ASEAN); they are zealously protective of the territorial waters of the Strait of Malacca; and they have the same ethnicity (Malay), language (bahasa Malayu/Indonesia), and religion (moderate Sunni Islam). Moreover, nationalism is prominent in both countries, although with one major difference. Indonesian nationalism is colored by a slightly anti-American hue; Malaysian nationalism has a markedly pro-American hue. In bygone years, the attitudes of the two countries were exactly reversed.

Economic relations between these two countries prosper and are likely to be spurred if and as ASEAN's plans for a regional free-trade zone are realized. However, there are similarities as well as differences in the structures of Indonesia's and Malaysia's economies, some of which may limit the benefits realized from expanded trade and investment between them.

For example, one difference in the structure of their economies is that Indonesia is a relatively large producer and exporter of crude oil and Malaysia imports oil. Thus, Indonesia benefits from higher oil prices while Malaysia benefits from lower ones. However, because of savings in transportation and other costs, both countries benefit from their increased bilateral trade in fossil fuels. Where their economies' product lines are similar (e.g., labor-intensive manufactures, rubber, copra), Indonesia and Malaysia tend to benefit more if they trade with third parties that have different product vectors rather than with each other. Currently, their trade with the rest of the world greatly exceeds their bilateral trade, as well as their intra-ASEAN trade.

Despite their differences, the two countries' preponderance of common interests and characteristics makes each country's decision about affiliating with PSI depend heavily, although not exclusively, on the other's decision. The probability that either one will affiliate with PSI is appreciably higher if the other has affiliated or signifies its intention to do so. It can be presumed that if either country were contemplating changing its stance on PSI, it would consult with the other before making the change.

We next consider the costs and benefits (disadvantages and advantages) of PSI affiliation for Indonesia and Malaysia separately and from each country's perspective. This sets the stage for attempting to distill from our assessment how the United States and like-minded PSI members might structure for each country an approach and an invitation that would enhance the prospects for affirmative decisions by both.

Indonesia

Indonesia has neither joined PSI nor participated in PSI exercises. While visiting Jakarta in March 2006, U.S. Secretary of State Condoleezza Rice invited Indonesia both to observe and participate in PSI activities. However, Indonesian officials continued to question the legality of PSI under the United Nations International Convention on the Law of the Sea (UNCLOS) and to worry that PSI members in the course of their PSI activities (especially their possible interdiction efforts) would infringe upon Indonesian sovereignty. On June 8, 2006 (two days after meeting with his U.S. counterpart, Donald Rumsfeld), Indonesia Defense Minister Juwono Sudarsono announced that Jakarta would consider endorsing PSI. However, he stipulated that Indonesia would participate only in an ad hoc manner and would not be active in all aspects of PSI.[1] He also pointed out that PSI participation would assist Jakarta in building its military capacity to patrol the Strait of Malacca.

In April 2007, the Indonesian Ministry of Defense stated that the Parliamentary Commission on Foreign Policy and Defense had, following discussion of the matter, strongly rejected participation in PSI, thus ending the government's attempt to subscribe formally to the PSI principles.[2]

---

[1] Michael R. Gordon, "Indonesian Scolds U.S. on Terrorism Fight," *New York Times*, June 7, 2006.

[2] Communication to the authors from Dr. Hasjim Djalal, Ambassador at Large and Senior Adviser on Maritime Affairs to the Indonesian Government, April 2007.

There are three salient and related aspects to consider in trying to understand Indonesia's assessment of the balance between advantages and disadvantages of PSI affiliation: sovereignty, law of the sea, and independent foreign policy:

Sovereignty. It is almost a truism that countries that have experienced long periods of colonial rule (in the case of Indonesia, by the Netherlands for two centuries; in the case of Malaysia, by Britain for nearly as long) tend to be highly sensitive to real or possible infringement of their sovereignty. Indonesia's sensitivity is especially acute because of its geographic size and dispersal—the country consists entirely of islands whose aggregate land mass (equal to more than three times the size of Germany) is interspersed by territorial and international waters coursing over a combined area of land and sea that is as large as the geographic area of the United States.

This sensitivity is evident in communications we have had with unofficial, informed, and influential contacts in Indonesia. As these and other sources in Indonesia see matters, PSI may entail

> initiation of interdiction . . . [of] suspected . . . national flag vessels . . . in international or national territorial waters . . . [that] would potentially interfere [with] Indonesia's territorial sovereignty . . . [by] internationalization of Indonesian territorial waters . . . and opening space within Indonesia's territory for external powers in their pursuit of WMD and other sensitive materials and technology.[3]

This quotation suggests a misunderstanding of PSI's principles and obligations, which we see as not only part of the problem of Indonesia's nonaffiliation, but also, as we discuss later, a clue to possible resolution of the problem. One potential resolution was suggested by Indonesia President (and former Army General) Yudhoyono following a visit to Jakarta in June 2006 by then-Secretary of Defense Rumsfeld: Indonesia would undertake "studies with regard to partial and ad hoc

---

[3] Memorandum to one of the authors from the Indonesian Center for Strategic and International Studies in Jakarta, April 2007.

implementation of some provisions of the PSI which may be adapted by Indonesia within the framework of Indonesia's sovereignty."[4]

Law of the sea. Separate from, but related to, its sensitivity about PSI's possible infringement of its sovereignty is Indonesia's questioning of PSI's legality under UNCLOS. Specifically, Indonesia has expressed the view that PSI commitments would violate UNCLOS's stipulation of the protected right of innocent passage.[5]

However, juxtaposed to the right of innocent passage is UN Security Council Resolution (UNSCR) 1540, adopted in April 2004, which allows—indeed, obligates—all states to "adopt and enforce appropriate effective laws which prohibit any non-state actor to manufacture, acquire, possess, develop, transport, transfer or use nuclear, chemical or biological weapons and their means of delivery."[6]

The juxtaposition of the UNCLOS "innocent passage" and UNSCR 1540 raises the issue of possible inconsistency between them. For example, what if there were credible intelligence for suspecting that a state-flagged vessel was transporting WMD/missile items to destinations or recipients that might be or might be suspected of being linked to a non-state actor, or that might be suspected of not being able to prevent a non-state actor from acquiring the transported cargo once it was off-loaded? Might not these circumstances provide grounds under UNSCR 1540 for suspending UNCLOS's specified right of innocent passage?

Our purpose in raising this question is to suggest that the right of innocent passage is not unqualified, and that UNSCR 1540 itself opens up possibilities and scenarios that qualify it. Moreover, UNCLOS (Article 19) explicitly introduces general qualifications to the right of innocent passage by plainly acknowledging that passage is not "innocent" if it is "prejudicial to the peace, good order or security of the coastal state," or is not "in conformity . . . with other rules of interna-

---

4 Juwono Sudarsono (Indonesia's Defense Minister), "U.S. Secretary of Defense," June 13, 2006.

5 "Indonesia Questions US Proposals on Proliferation Security Initiative," *Jakarta Post*, March 16, 2006.

6 UNSCR 1540 (2004), S/RES/1540 (2004), April 28, 2004.

tional law." That there is ample room for different interpretations of several nouns and adjectives in this quotation highlights the qualifications that may be associated with "innocent passage." For example, it may be justifiable to interpret the illegitimate transport of WMD/missile items as prejudicial to peace or security and therefore as voiding the right of innocent passage.

Independent foreign policy. Because of Indonesia's colonialist history, zealous concern for protecting its sovereignty, and status as the world's largest Muslim country, it views with pride its role as an advocate and leader of countries that profess nonalignment and independence in their foreign policies. In the Indonesian context, this means *nonalignment with* and *independence of* the United States or, at the least, circumspection in undertaking or appearing to undertake obligations that might compromise a country's independence as the result of an excessively close link to the United States[7]

Notwithstanding the resumption of military-to-military relations between Indonesia and the United States, nor Indonesia's pervasive admiration for the U.S. economy and many aspects of its society and culture, the ubiquity of the U.S. economic and military presence in international affairs makes Indonesia especially cautious about too close an embrace, lest it compromise or appear to compromise Indonesian "independence" and "nonalignment."

The idea that PSI affiliation would entail such a compromise, or constitute too close an embrace of the United States, suggests some misunderstanding in Indonesia about PSI's purpose and scope. The consistency between deliberate and assertive independence of the United States, on the one hand, and PSI affiliation, on the other, is underscored by the examples of France (during Jacques Chirac's presidency) and Russia (under former President, now Prime Minister Vladimir Putin)—two countries that are especially assertive about their independence but nonetheless are actively participating members of PSI.

---

[7] For a discussion of nonalignment and independent foreign policy, see Nina Hachigian and Mona Sutphen, *The Next American Century*, 2008, pp. 142–145. These pages mainly focus on India, but much of what is stated applies as well to Indonesia's view of nonalignment. See also Angel Rabasa and Peter Chalk, *Indonesia's Transformation and the Stability of Southeast Asia*, MG-1344, Santa Monica, Calif.: RAND Corporation, 2001.

The French and Russian memberships, as well as the multinational leadership of PSI's OEG, provide clues about how to remedy Indonesia's misunderstanding and to encourage its affiliation. We directly address this topic later, in Chapter Three, where we consider ways of encouraging each of the five countries to reassess its current stance on PSI affiliation.

Malaysia

Compared with Indonesia, Malaysia is more supportive of PSI and has participated as an observer in PSI exercises, three in all. In April 2007, Malaysia Defense Minister and Deputy Prime Minister Razak told reporters that Malaysia supported the principles of PSI but was still studying its legality and had no immediate plans to join. He also said: "[N]evertheless, we've some cooperation with ASEAN countries to share information and intercept ships carrying suspicious cargo."[8]

Two of the three aspects related to Indonesia's assessment of whether to affiliate with PSI—i.e., sovereignty and law of the sea—also apply, in much the same terms, to Malaysia. The third—independent foreign policy—has a different, more complex, and generally more favorable tone when applied to Malaysia.

Although Malaysia views itself and is viewed by others as identifying with nonaligned foreign policy, its orientation is tilted in the reverse direction from that of Indonesia. Rather than being marked by a touch of anti-Americanism, Malaysia's current Prime Minister, Abdullah Badawi, is vocally pro-American and pro-Western. Indeed, he has called for "strategic partnership" between the West and the Muslim world, and he regularly and frequently emphasizes the common interests of Malaysia and the United States in the joint struggle against terrorism.[9]

This emphasis may be a legacy of Malaysia's protracted and bitter experience with insurgency during the 1950s and 1960s, in contrast

---

[8] "Malaysia Still Studying Membership in PSI, Says Najib," *Malaysian National News Agency*, April 17, 2007. See also Stephanie Lieggi, "Proliferation Security Initiative Exercise Hosted by Japan Shows Growing Interest in Asia but No Sea Change in Key Outsider States," *WMD Insights*, December 2007–January 2008.

[9] "Security, Peace and Prosperity for All," *Reuters*, February 25, 2008.

to Indonesia's less intense and shorter-lived experience with terrorist bombings in Bali in 2002. Or it may be one way that Badawi has chosen to distance himself from his often virulently anti-American predecessor, Prime Minister Mahathir Mohammad. In any event, Malaysia's expressions of common interest with the United States in cooperative efforts to combat terrorism, as well as in close bilateral relations with the United States more generally, are unequivocal and yet do not preclude Malaysia's continued profession of nonalignment and foreign policy independence.[10]

In 2004, Malaysia placed two of its ports, Port Klang and Tanjung Pelepas, under the Container Security Initiative (CSI), which was launched by the U.S. Customs and Border Protection (CBP) as part of America's anti-terrorism program. Indonesia, in contrast, has no ports participating in CSI. American CBP officers are stationed at foreign CSI ports to observe the security screening conducted by host-country officials. Of course, the CBP officers focus only on cargo containers destined for or transiting through the United States. Nevertheless, Malaysia's participation in CSI signifies its lesser sensitivity (relative to that of Indonesia) about foreign cooperation in efforts to combat terrorism (especially concerning terrorists' possible attempts to gain access to WMD), and hence its higher probability (compared with that of Indonesia) of affiliating with PSI.

There are several other positive indicators of prospects for Malaysian affiliation with PSI. One of these is its attendance at three PSI exercises, most recently in October 2007.[11] Another is Malaysia's increased efforts, through its modernized and expanded Coast Guard (Malaysian Maritime Enforcement Agency), to patrol and monitor the Malacca Strait jointly with Thailand, Singapore, and Indonesia. Still another positive sign is an April 2007 statement by Malaysia's foreign minister that "[e]ven if we are not signatory [to PSI], there are instances

[10] The sharp change in the current Malaysian government's stance toward cooperation with the United States from that of the preceding, Mahathir government suggests, of course, that the stance might change again in the future.

[11] Ministry of Foreign Affairs of Japan, *PSI Maritime Interdiction Exercise "Pacific Shield 07," Hosted by the Government of Japan (Overview and Evaluation)*, October 18, 2007.

in which we can cooperate."[12] Finally, affiliation with PSI, as well as CSI, can perhaps contribute to sustaining or even enhancing Malaysia's competitive position in international commerce, whereas nonaffiliation may detract from that position. This possible benefit might arise to the extent that trading partners and their insurers, at present inclined to favor entrepot transactions through Singapore—a PSI participant—may accord equal treatment to Malaysia.

There are also several negative indicators. One is the fact that during the tenure of its preceding prime minister, Mahathir, Malaysia was implicated in the proliferation network of Pakistan's A. Q. Khan. As part of that network, Malaysia had a key factory that produced centrifuge components capable of enriching uranium gas to provide weapons-grade fissionable material. One of the main owners of the factory was the current prime minister's son.[13] In response to U.S. pressure and after initial resistance, however, Malaysia did agree to cooperate with the International Atomic Energy Agency (IAEA) in its investigation, including inspection of the culpable factory, and to sign the additional protocol to its nuclear safeguards agreement with IAEA. Indeed, Malaysia may now have an extra incentive to demonstrate its commitment to nonproliferation of WMD/missile items, possibly enhancing its receptivity to PSI affiliation.

A second, not unrelated, negative consideration is Malaysia's record of support for Iran in its dispute with the West over Iran's nuclear program.[14] Further, Iran's recent award of a large oil contract to a Malaysian firm may cause Malaysia to view Iran as an important and increasingly valuable business connection.[15] Malaysia therefore may not be inclined to view Iran as a state of "proliferation concern," which would reinforce its reluctance to affiliate with PSI.

---

12 "Malaysia in No Hurry to Join US-led Security Pact," *Reuters*, April 17, 2007.

13 "Malaysia Arrests Alleged Black Market Nuclear Agent," *Taipei Times*, May 30, 2004.

14 James Martin Center for Nonproliferation Studies, "NAM Chair Malaysia Skeptical of UNSC Involvement," January 2006.

15 "Iran, Malaysia Sign $16b Oil Deal," *China Daily*, December 27, 2007.

Pakistan and India

Our reasons for grouping India and Pakistan together for our assessment are a complex mix of strategic, political, financial, and technological interactions. Both countries already have nuclear weapons and the technical capability to reproduce and export them. Further, the longstanding tension between the two makes neither disposed to endorse PSI without considering what the effect on the other might be. For example, if Saudi Arabia, which recently endorsed PSI, were in the future to seek a "Sunni bomb" to counter Iran's acquisition of a "Shiite bomb," either or both India and Pakistan might wish to cooperate with Saudi Arabia in this endeavor and might anticipate that endorsement of PSI would inhibit such cooperation.

Pakistan

We found no official Pakistani statement expressing views on Pakistan's affiliation with PSI. This silence is not surprising given that the government first has to deal with the aftermath of the A. Q. Khan transnational nuclear-smuggling network brought to light by Khan's public confession on February 2, 2004. However, like Malaysia, Pakistan has been an observer in three PSI exercises: Deep Sabre, hosted by Singapore in 2005; Pacific Protector, hosted by Australia in 2006; and Pacific Shield, hosted by Japan in 2007.

As a financially challenged country and one with the capabilities to supply nuclear weapons, Pakistan might have commercial and political reasons for wanting to be considered a potential supplier in a scenario such as the Saudi Arabian one mentioned above. And if claims that Saudi Arabia has, in the past, contributed financially to the development of Pakistan's own nuclear capability are true, they may add political credibility to this reasoning.[16] It is also worth noting that

---

[16] Steve Weissman and Herbert Krosney, *The Islamic Bomb*, Times Books, New York, 1981. See also "Israel: Saudi Arabia's Purchase of Nuclear Warheads Said 'Threat to World Peace,'" translated excerpt, *Foreign Broadcast Information Service*, October 22, 2003. As reported in the article, Knesset member (MK) Yuval Steinitz, chairman of the Knesset Foreign Affairs and Defense Committee, said, "Dr. Patrick Clawson, director of research at The Washington

Pakistan's predominantly Sunni Muslim adherence adds further plausibility to this scenario.

As suggested earlier, were a Saudi disposition toward WMD acquisition to emerge, Pakistan might perceive PSI affiliation as entailing limitations and restrictions on its pursuit of the role of supplier, either covertly or overtly. Thus, Pakistani leaders might see PSI affiliation as disadvantageous for both political and commercial reasons.

However, several indicators suggest that Pakistan may have a more favorable view of PSI affiliation. One such indicator is Pakistan's official report on implementation of UNSCR 1540, which was submitted on October 27, 2004, to the UN Security Council committee established pursuant to the resolution.[17] Indeed, the principles, affirmations, legislation, and administrative measures described in the report are closely congruent with the basic principles constituting PSI's ground rules.

Other positive indicators include Pakistan's participation in the Secure Freight Initiative (SFI), although Pakistan has thus far abstained from participating in CSI. SFI, which is complementary to PSI, is designed to improve detection of nuclear and radiological materials packaged in maritime containers and thus to raise the probability of interdicting such materials.[18]

On June 9, 2007, a Pakistani Foreign Ministry spokesperson announced that Pakistan would join the Global Initiative to Combat

---

Institute for Near East Policy and a leading US expert on the prevention of nuclear proliferation, told us he believes it was Saudi Arabia that financed the Pakistani nuclear program."

[17] *Pakistan's National Report on National Measures on the Implementation of Security Council Resolution 1540* (2004), United Nations S/AC.44/2004/(02)/22, October 27, 2004. A summary of this 2004 report is provided in Appendix C.

[18] SFI is sponsored by the U.S. departments of Homeland Security (DHS) and Energy, which provide funding for the acquisition and installation of radiation-detection equipment in foreign ports to detect nuclear and radiological materials in maritime containers before they are sent to the United States. Under SFI, the containers bound for the United States in these foreign ports will be scanned and the gathered data, including a detection alarm, will be sent in near-real-time to CBP agents working in these foreign ports, the DHS National Targeting Center in Virginia, and the host country officials. The United States has been working with host governments to establish protocols for a quick resolution by the host government, including the instruction to carriers not to load the container until the risk is resolved.

Nuclear Terrorism led by the United States and Russia.[19] The announcement affirmed that Pakistan's "legislative, regulatory and administrative infrastructure" had recently been improved and that Pakistan was prepared to address the threat of nuclear and radiological terrorism. This perhaps suggests that an invitation to Pakistan to become affiliated with PSI might be more convincing if it came jointly from Russia and the United States. Indeed, the invitation might be most effective if it came from all of the nuclear-weapons participants in PSI—France, the United Kingdom, Russia, and the United States—thus signaling both elevation and recognition of Pakistan's status as a nuclear weapon state.

Even the previously mentioned negative factors might be used to motivate Pakistan to affiliate with PSI. The A. Q. Khan nuclear-smuggling network sent nuclear technology and equipment to third parties, including North Korea, Iran, Libya, and perhaps Saudi Arabia. Pakistan's affiliation with PSI might be viewed as a way for Pakistan to redeem the nonproliferation setback caused by the Khan network's activities.

Finally, Pakistan's participation as an observer in three PSI exercises provides grounds for reasonable optimism about its current prospects for PSI affiliation.

India

Defense Minister Pranab Mukherjee stated in June 2006 that he disfavored India's participation in the "U.S.-led" PSI.[20] However, India participated as an observer in three PSI exercises: Pacific Protector, hosted by Australia in 2006, Pacific Shield, hosted by Japan in 2007, and Maru, hosted by New Zealand in 2008. Also, in December 2004, India hosted a seminar on maritime security at which India and Sri Lanka

---

[19] "Pakistan Joins Initiative to Combat Nuclear Terrorism, Establishes Strategic Export Control Division," *International Export Control Observer*, 11, June/July 2007, p. 2. The Global Initiative to Combat Nuclear Terrorism, announced by the U.S. and Russian presidents on July 15, 2006, aims to prevent terrorist access to nuclear weapons.

[20] P. S. Suryanarayana, "The Country Is Enhancing Its Defence Preparedness, Says Minister," *The Hindu*, June 4, 2006.

discussed a common law to combat maritime terrorism by preventing illegal shipments of weapons from entities in India to Sri Lanka.[21]

We start by considering India's internal political situation, which in the past provided formidable obstacles to PSI affiliation but has recently changed in ways that appear to make India's participation more promising. We also consider external factors relevant to India's decision to affiliate, such as the remote possibility of Saudi Arabia seeking a nuclear capability, or the less remote possibility of Iran doing so, with India perhaps wishing to retain maneuverability as a possible alternative supplier of components. However, these external considerations are assuredly only part of the explanation for India's nonaffiliation; the internal political dynamic in India most likely plays the dominant role.

Until recently, the Communist Party of India (Marxist), or CPI(M), and its leftist allies held the key swing votes supporting the continuation of the Congress Party's governing United Progressive Alliance (UPA). The CPI(M), which is vigorously anti-American, had become increasingly concerned about the growing ties between the United States and India. This concern came to a head over the proposed nuclear cooperation agreement between the two countries.[22] The result was prolonged political turmoil, with the CPI(M) threatening to bring down the UPA government.

Given this political situation, the Indian government was reluctant to move ahead with other agreements with the United States as long as the issue of the nuclear cooperation agreement remained unresolved. For example, India decided not to sign a logistics support agreement with the U.S. in early 2008.[23] And with regard to India participating in PSI specifically, the situation was further complicated by the

---

[21] Per source cited ("Meet Focuses on Maritime Security," *The Hindu*, December 19, 2004) in "Indian Government Boosts Maritime Security; Stays Cool to Proliferation Security Initiative," *Asian Export Control Observer*, 6, February/March 2005, p. 9.

[22] Formally, "Agreement for Cooperation Between the Government of the United States of America and the Government of India Concerning Peaceful Uses of Nuclear Energy (123 Agreement)," August 3, 2007.

[23] "India Says No to Logistics Deal with U.S., for Now," *The Times of India*, February 27, 2008.

fact that PSI is referenced in the Hyde Act—the enabling legislation for the nuclear agreement.[24] The Hyde Act states a number of policies to be followed in South Asia that involve securing India's support for various U.S. policy objects. In particular, it states that U.S. policy in South Asia should include securing India's "full participation in the Proliferation Security Initiative" and "formal commitment to the Statement of Interdiction Principles of such Initiative."

In an open letter (dated September 8, 2007) to members of India's parliament, the CPI(M) stated: "The [Hyde] Act concerns itself with areas outside nuclear cooperation and contains objectionable clauses to get India to accept the strategic goals of the United States"[25] and that these include "Indian participation and formal declaration of support for the US's highly controversial Proliferation Security Initiative including the illegal policy of interdiction of vessels in international waters."

It was thus clear, at this point, that as long as the CPI(M) maintained its key role in keeping the UPA in power, India was unlikely to be able to participate formally in PSI. However, in June 2008, as the November 2008 U.S. presidential election grew more imminent, the Indian government deemed it important to conclude the nuclear cooperation agreement with the United States and decided to move ahead with it. In response to this action on the part of the government, the CPI(M) withdrew its support from the UPA. In the resulting no-confidence vote in July 2008, the UPA succeeded in getting the support of new allies and won the vote. Now, with the CPI(M)'s objections no longer relevant, India may find it possible to formally participate in PSI.

As for the external factors that may influence India's disposition toward PSI, there is the remote scenario, mentioned earlier, in which Saudi Arabia seeks its own WMD and India has some appeal simply as an additional source (besides Pakistan) of WMD technology, provid-

[24] Formally, *The Henry J. Hyde United States–India Peaceful Atomic Energy Cooperation Act of 2006*, Public Law 109-401, 109th Congress, 1st Sess., December 18, 2006.

[25] "CPI(M)'s Open Letter to Members of Parliament," *People's Democracy (Weekly Organ of the Communist Party of India [Marxist])*, Vol. XXXI, No. 38, September 23, 2007.

ing buyer-beneficial price competition (the Saudi royals are well known for business acumen and price sensitivity). India might also provide a Saudi buyer with a secondary political ally in South Asia, an ally against a Saudi-envisioned future that might include an aggressively inclined Shiite Iran, and perhaps Iraq as well.

From India's viewpoint, including its continuing concern over Pakistan's instability and the effect this will have on stability in the subcontinent, there may be merit in deferring affiliation with PSI as long as Pakistan remains unaffiliated. Once again, this judgment presumes that India might see its affiliation as implicitly restricting its freedom to maneuver if a Saudi Arabian or Iranian interest in WMD acquisition were to emerge.

Uncertainty about what constitutes countries of "proliferation concern" in the PSI lexicon, and about how and by whom this designation will be made, provides plausible grounds for Pakistan and perhaps India to have reservations about PSI affiliation. India's stance on nonaffiliation may be significantly affected by that of Pakistan. For example, if nonaffiliation is seen as offering greater maneuvering freedom in the event that a Saudi or an Iranian option were to emerge or, conversely, if affiliation is construed as entailing more constraints on this freedom, India's reasons for nonaffiliation would be understandable.

China

China's Position on PSI

Treating China as a "group of one" is, from the standpoint of PSI affiliation, warranted on several grounds. In addition to being a nuclear weapons state, China is the second or third largest economy in the world (depending on whether one uses, respectively, purchasing power parity or nominal exchange rates to convert gross domestic product [GDP] from yuan to dollars), the fourth or fifth largest global trading country, and the third or fourth largest global weapons exporter—perhaps involving weapons exports that might inadvertently include WMD components. By virtue of these, and other, attributes, China shares a mix of political, economic, and security interests and transactions with the two major current and prospective sources of "proliferation concern": North Korea and Iran.

Of the five countries, China has most frequently articulated its position on PSI. Several statements issued by China's Foreign Ministry are germane. In December 2003, for example, a spokesperson from the Chinese Foreign Ministry stated:

> China understands the concerns of the PSI participating countries over the proliferation of WMD and their delivery systems. However, there are also many concerns in the international community about the legitimacy and effectiveness of PSI interdictions and consequences that may arise there from. The PSI participants should take this into serious consideration. China consistently holds the view that proliferation issues should be resolved within the international legal frameworks by political and diplomatic means and that any nonproliferation measures to be taken should serve to promote international and regional peace, security and stability.[26]

In October 2004, a China Foreign Ministry spokesperson expressed concern that the PSI might allow "military interception, which is beyond the limits of international law."[27] The main issue, however, has been the legitimacy and consequences of interdiction. China is concerned that PSI could infringe upon the right of innocent passage of Chinese ships through the territorial waters of PSI members, such as the Straits of Malacca. This concern shows up in a Foreign Ministry statement about PSI made in May 2007:

> China is firmly opposed to proliferation of WMD and their means of delivery and stands for the attainment of the non-proliferation goal through political and diplomatic means. We understand the concern of PSI participants over the proliferation of WMD and their means of delivery and share the non-proliferation goal of the

[26] Ministry of Foreign Affairs of the People's Republic of China, "Foreign Ministry Spokesperson's Press Conference on 4 December 2003," December 4, 2003.

[27] Ministry of Foreign Affairs of the People's Republic of China, "Foreign Ministry Spokesperson Zhang Qiyue's Press Conference on 26 October 2004," October 27, 2004.

> PSI. China supports the cooperation among the PSI participants within the framework of international law. However, the international community including China remains concerned about the possibility that the interdiction activities taken by PSI participants move beyond the international law.
>
> China, pursuant to its non-proliferation policy, national laws and international commitments, has conducted fruitful cooperation with the international community including some PSI participants in non-proliferation field in recent years. We are ready to continue and further strengthen such constructive exchange and cooperation.[28]

This Chinese concern that PSI members might interdict in situations that China considers to be innocent passage is the same one we discussed earlier in connection with Indonesia's reluctance to affiliate with PSI.

While China has not participated as an observer in any PSI exercises, it has placed three of its ports under CSI: Hong Kong, on May 5, 2003; Shanghai, on April 28, 2005; and Shenzhen, on June 24, 2005.

Costs and Benefits of China's PSI Affiliation

The multiple and complex interests that China has at stake in connection with PSI result in a mix of incentives and disincentives for formal affiliation. For example, China's regional security interests have motivated it to host and restart the Six-Party talks on North Korea and to cooperate closely with the United States in efforts to halt and reverse North Korea's nuclear programs. Nevertheless, China's leaders think (as do Russia's leaders) that their nation's relations with North Korea, and its ability to influence North Korean behavior, are enhanced by avoiding formal commitment to PSI, which might look threatening to North Korea. In line with this thinking, China views blandness and "carrots" as preferable to coercion and "sticks" for dealing with North Korea, which is different from the U.S. approach. The fact that South

[28] Ministry of Foreign Affairs of the People's Republic of China, "Proliferation Security Initiative," May 21, 2007.

Korea used to hold a view similar to China's and has nonetheless joined several PSI exercises as an observer (though it has not yet become a PSI participant) may suggest a path for Chinese affiliation with PSI. For example, endorsement of PSI by the new administration in South Korea's Blue House might be linked to a similar action by China.

Toward Iran, China's approach is similar to its approach to North Korea: It employs temporizing and carrots rather than pressure and sticks—again, a different approach from that of the United States. Despite this difference in the two countries' approaches and China's currently substantial and prospectively growing commercial interests in Iran, however, China has become a full participant in the Group of Six (joining the United States, Russia, the United Kingdom, France, and Germany) in supporting (mild) Security Council sanctions against Iran.

Thus, China has distinctive attitudes toward Iran and North Korea, the two states of primary "proliferation concern." Yet China is also troubled by an internal separatist/terrorist movement among the Turkic-speaking Islamic Uighurs in Xinjiang, who might conceivably seek nuclear weapons or other WMD in furtherance of what China refers to as the movement's "splittist" objectives.[29] Consequently, China might be expected to choose PSI affiliation as a way to minimize the risk of Uighurs accessing WMD. However, China's leaders may instead decide that they can minimize this risk more effectively and aggressively by unilaterally acting to intercept suspicious cargo from North Korea. This would avoid the implicit obligations that PSI affiliation may entail, and would not oblige China to employ blander interception methods than it might use unilaterally.

Another consideration that may affect China's reluctance to affiliate with PSI is the possibility—remote, but not inconceivable—that China might be willing to accede to a future Saudi request to acquire a "Sunni bomb" if the Shiites were to acquire the precursor weapon. Although such a course of action is unlikely, it may become more plausible in some situations—for example, China might

[29] Conversations between one of the authors of this report and Chinese experts, in Beijing, May 2008.

cooperate with Pakistan in meeting the Saudi request by providing a part of the WMD system, such as the delivery component. In fact, in the late 1980s, China provided the Saudis with CSS-2 intermediate-range ballistic missiles, which China had designed to carry nuclear warheads for its own nuclear arsenal.[30] Saudi Arabia assured the United States that it had no intention of equipping them with nuclear warheads. However, should the Saudis decide to acquire a nuclear capability, the Chinese could enable the existing and new Saudi missiles to carry nuclear warheads. China thus might see PSI affiliation as a restriction on its latitude in a situation that might involve the Saudis seeking WMD systems. But, then, were China to consider assisting Saudi Arabia (or any country, for that matter) with its acquisition of WMD/missile items, the more binding concern probably would be violation of the Nuclear Nonproliferation Treaty (NPT), to which China acceded in 1992, and the Missile Technology Control Regime, which China accepted in 2000.

China also has substantial economic interests in Iran relating to oil and other bilateral investment and trading opportunities, and China's leaders may regard these interests as possibly compromised by affiliation with PSI. For example, China has sought, and perhaps has already concluded, long-term contracts for future imports of Iranian oil and may worry that these might be imperiled by PSI affiliation.[31] Additionally, China is making substantial investments in Iran—notably, multibillion-dollar investments in a refining capacity designed to enable Iran to export gasoline in three or four years (Iran currently imports 40 percent of its gasoline consumption). Not implausibly, China may anticipate that these investments would be compromised by PSI affiliation.

Another complexity arises in connection with what China views as an awkward constraint being imposed by the *U.S. National Defense Authorization Act for Fiscal Year 2000* (PL 106-65), which precludes

---

[30] King Fahd pledged that Saudi Arabia would not arm the missiles with unconventional warheads and would not use them in a first-strike mode. Youssef Ibrahim, "Saudis Warn They May Use Missiles Against Iran," *New York Times*, April 29, 1988.

[31] "China, Iran Sign Biggest Oil and Gas Deal," *China Daily*, October 31, 2004.

"inappropriate exposure" by U.S. military-to-military contacts with China that might involve "advanced logistical operations, . . . surveillance and reconnaissance operations, . . . [and] other advanced capabilities of the Armed Forces."[32] This lingering restriction is part of the long-standing U.S. effort to induce China to be more transparent in its military budgeting, research and development, and other military activities in order to approximate a quid pro quo basis in China-U.S. military-to-military exchanges. The logic behind releasing PL 106-65 is as follows: If China is not transparent in various military information, the United States will not be transparent in discussing its advanced capabilities in the military-to-military contacts. However, PL 106-65 does not restrict search-and-rescue or humanitarian operations in concert with the People's Liberation Army (PLA), and, indeed, search-and-rescue exercises (SAREXs) have taken place. Nevertheless, it is entirely possible that China is concerned lest participation in PSI collide with what it views as demeaning prohibitions contained in PL 106-65, about which China is sensitive and resentful. For example, China's leaders may be concerned that China might be abruptly and awkwardly asked to leave a PSI interdiction exercise or operation so that advanced U.S. detection capabilities would not be "inappropriately exposed" to the Chinese.

In seeking to ease this concern and, more generally, to provide positive incentives for China's participation in PSI, the United States might consider legislative or other means of loosening these PL 106-65 restrictions, perhaps in exchange for China taking specific steps toward increased military transparency.

China's Reluctance to Legitimize PSI Interdictions

The most controversial element of PSI is the act of interdiction. It is logical to think that the UN can be called on to lend support to PSI in general and to legitimize interdictions of ships suspected of delivering WMD/missile items to state and non-state actors of proliferation concern. The United States attempted this route with UNSCR

[32] *National Defense Authorization Act for Fiscal Year 2000*, Public Law 106-65, 106th Congress, October 5, 1999.

1540, which was adopted unanimously by the Security Council on April 28, 2004. Originally, the United States wanted to use the UN to change international law and criminalize WMD proliferation activities in order to support PSI. China and Russia opposed this move and threatened to veto any resolution that endorsed PSI.[33] Moreover, the final text of UNSCR 1540 was agreed to only after the United States accepted China's demand (and a veto threat) to drop a provision specifically authorizing the interdiction of vessels suspected of carrying WMD.[34]

---

[33] William Hawkins, "Chinese Realpolitik and the Proliferation Security Initiative," Association for Asia Research, February 18, 2005.

[34] Mark Valencia, "The Proliferation Security Initiative: Making Waves in Asia," *Adelphi Papers*, 45:376, 2005, p. 48.

CHAPTER THREE

Cutting the Gordian Knot: Principles and Measures to Encourage PSI Affiliation

PSI Benefits

As evidenced by statements in their reports to the UN on national measures to implement UNSCR 1540, as well as their signing of various international treaties and agreements (see Appendix C), all five of the subject countries—Malaysia, Indonesia, India, Pakistan, and China—have committed to nonproliferation. PSI allows affiliated countries to join in preventing "the flow of these items [WMD, their delivery systems, and related materials] to and from states and non-state actors of proliferation concern."[1] Through intelligence sharing and cooperation in interdiction, a PSI participant is more likely to learn that one of its flagged ships is carrying illicit WMD items; it is also more likely to be able to obtain immediate help in inspecting cargo beyond its reach. Additionally, PSI exercises help improve participants' interdiction procedures and capabilities.

PSI affiliation also offers the qualitative benefit of contributing to more-cooperative strategic relations with other PSI participants. For example, Saudi Arabia and Singapore may plausibly view endorsement of PSI as having a positive effect on their overall strategic relations with the United States. Any of the five countries we assessed is likely to consider this benefit of PSI affiliation while at the same time weighing the

[1] The quoted material is from *The Proliferation Security Initiative (PSI)* (U.S. Department of State, Bureau of International Security and Nonproliferation, 2008) (replicated in Appendix A).

possible costs of closer alliance with certain PSI participants, notably including the United States. The five principles we discuss below can reduce these costs while sustaining the benefit.

In addition, each country, depending on its situation, may plausibly consider other potential benefits. Each of the five has been improving its inspection capabilities for imports and exports, as well as the domestic laws controlling them. Participation in CSI, SFI, and the Global Initiative to Combat Nuclear Terrorism can contribute to such improvements, as does taking part in the Export Control and Related Border Security Assistance Program (EXBS). PSI participation also includes workshops, training, and technical assistance to help countries improve import and export controls. While participation in these several initiatives and programs is not contingent on joining PSI, affiliation with them would interact synergistically to improve national inspection capabilities. Furthermore, PSI affiliation may increase access to inspection/interdiction equipment, patrol boats, and/or technical assistance from other PSI participants, although no instances of such access have occurred thus far.[2]

In the calculus of benefits from PSI affiliation, two types of benefits can be highlighted for the five countries' reconsideration. The first is enhanced *deterrence*—that is, increased risk to potential WMD proliferators that WMD/missile items will be intercepted, thereby discouraging proliferation and/or raising the costs of attempted proliferation. While deterrence is largely a collective, or "public good," benefit rather than a narrower, national one, it may have national appeal as a country's contribution to global security, or for a country (South Korea

[2] For example, although Indonesia has not yet joined PSI, the Japanese government offered Jakarta three midsized patrol boats (20 meters in length and costing U.S. $6.5 million each) to help enhance the effectiveness of Indonesian patrols. Such benefits could be expected to increase if Indonesia were to join PSI. This was the first time Japan offered equipment of this type to a developing country free of charge. This action helps Indonesia to maintain security on the straits through which more than 90 percent of Japan's oil passes. Tsuyoshi Nojima, "Japan to Offer Patrol Ships to Curb Piracy," *Asahi Shimbun*, March 17, 2005, in Lexis-Nexis Academic Universe, as cited in *Asian Export Observer*, 7, April/May 2005, p. 8.

may be an example) acutely concerned that proliferation could directly affect its security.

PSI affiliation might also yield direct economic benefits to participants through fewer and faster cargo inspections, resulting in reduced costs of delay. However, this type of benefit is unlikely to materialize unless purposeful government actions accompany affiliation with PSI.

The Five Principles

In Chapter One, we reviewed the interests and perspectives of the five countries that constitute, for them, plausible reasons for not affiliating with PSI. We now turn to the key policy question of this research: What measures, policies, and approaches can be invoked by the United States and other PSI affiliates that are likely to induce each of the five countries to lower its estimates of the costs (disadvantages) of affiliation and/or raise its estimates of the benefits (advantages) enough that it arrives at a positive (rather than negative or neutral) bottom line for affiliation? Implicit in this question is the premise that PSI's effectiveness will be enhanced through an increase in the number of participating nations, notably the five hold-out countries with which we are concerned.

To assist in answering our key policy question for each of the five countries, we set out five general principles, each one applicable to at least one of the five countries, and several jointly applicable to more than one country. These principles are intended to guide PSI actions related to the five countries. As such, they require the clarification and elaboration of PSI declaratory policies. The principles are as follows:

1. *Exercise U.S. leadership by ceding it to other PSI participants.* In some cases—Indonesia is one and, to a lesser but still important degree, so are India and China—affiliation and closer cooperation with the PSI "activity" are more likely to be forthcoming if an invitation is proffered by a PSI participant or participants other than the United States. The underlying reasons for this are different in the respective cases. Indonesia, for instance, has

a relatively "nonaligned" foreign policy and thus is less likely to be perceived as being infringed upon if principle 1 is invoked. China, with its (restrictive) leadership of the Shanghai Cooperation Organization (SCO), is more likely to appear implicitly challenged if a PSI invitation emanates unilaterally from the United States rather than as a shared responsibility through principle 1. Validation of principle 1 will require further accordance of more prominence and influence to PSI participants other than the United States.

2. *Interpret and apply "innocent passage" consistent with each state's own national legal authorization and its obligations under international law.* We suggested earlier that the right of innocent passage is both ambiguous and hermeneutical.[3] Thus, the intent of principle 2 is to emphasize that PSI affiliation does not further circumscribe an affiliate's latitude for interpreting and applying the right of innocent passage—whether by action or inaction.
3. *Affirm the validity of "territorial waters" and emphasize the locus of responsibility in the littoral countries.* This is truly a codicil to and elaboration of principle 2. Principal 3 in large measure reiterates most of principle 2, but it differs in emphasizing the perspective of the littoral countries rather than that of the flag states of ships that make the passage. The reason for including this principal is to provide more-direct evidence that PSI in general and the OEG in particular are aware of and sensitive to the special concerns of littoral states—notably, Indonesia, Malaysia, and China.
4. *Present PSI affiliation as incremental to agreements/commitments already arrived at.* All five countries have to some extent and to varying degrees made prior commitments and endorsed other agreements that, although separate from PSI, are consonant with PSI's goals and atmospherics. The details and variety of these commitments and agreements are summarized in Appendix C

[3] See the discussion of law of the sea in Chapter One's section on Indonesia.

and, for Pakistan, Appendix D.[4] The purpose of this principle is to suggest that for some of the countries, PSI affiliation might be eased by highlighting its modest, incremental character.

5. *Confer membership in the OEG,* ab initio. The rationale here is simply that the stature and importance of several of the five countries make it highly appropriate that their affiliation with PSI carry with it membership in the OEG—the group that provides PSI's leadership and manages PSI activities. Of course, OEG participation should be presented as an optional incentive rather than a mandatory obligation. Indeed, some of the countries may prefer not to be immediately engaged in the OEG even if willing to affiliate with PSI.

Note that our five principles do not include carrots and sticks related to issues outside PSI (such as peaceful nuclear assistance, including nuclear power plants for electricity generation). Instead, they focus on assuaging the concerns of the five countries (as well as any other unaffiliated countries) about PSI affiliation interfering with their existing international obligations and rights so that these countries will be able to more favorably re-estimate the costs and benefits of joining PSI.

Our objective is for the five to join PSI because they consider the benefits of participation to outweigh the costs—not because they want to use participation as a bargaining chip for obtaining benefits or avoiding penalties on issues unrelated to PSI or nonproliferation. Affiliation with PSI for its own sake is the goal, because it would make them more-active participants and, in the long run, meet the nonproliferation objective far better than affiliation for other reasons would. Moreover, offering these countries any special inducements for joining would set in motion most-favored-nation clamor from many of the 91 countries that have already joined.

Almost all of the five principles were implicitly or explicitly envisaged in the earlier discussion of the five countries' separate interests and

---

[4] A short description of various agreements will be included in a separate document prepared under this study.

perspectives. However, before applying the principles and the subsets of the principals to the five countries, further explication is useful.

Principle 1

The rationale for the apparent paradox of the first principle—exercise leadership by ceding it—was inherent in our discussions of Indonesia, India, Pakistan, and, to a lesser extent, China. The ubiquity, occasional heavy handedness, and sometimes dominance of the U.S. international presence may heighten a country's sensitivity and resistance to an invitation to affiliate with PSI that is extended directly by the United States. If the invitation is instead issued from and conducted by a non-U.S. PSI member or members, the reaction (especially, for example, in the case of Indonesia, which is a professedly "nonaligned" country) may have a greater likelihood of being positive. The additional offer of immediate participation in the OEG—a body that acts through consensus rather than the decision of any single member—would give added weight to this point.

Principles 2 and 3

As noted earlier, interpretation of the right of innocent passage (principle 2) is rife with ambiguity. This point is of concern to some of the five countries as a matter of international law (e.g., China and India) and is of special concern to the littoral states of Indonesia and Malaysia. (Principle 3's special concern to the latter two countries is why we describe it as a codicil to principle 2.) For both of these principles, we suggest that a positive response to an invitation to affiliate with PSI is more likely to be forthcoming if the locus of responsibility for interpreting, applying, or waiving innocent passage is conveyed to the five countries as being clearly lodged in the proximate state rather than in other PSI affiliates. In the formal statement of PSI's interdiction principles (see Appendix B), PSI's own enabling principles all emphasize that their interpretation and application are heavily vested in the individual, sovereign members of PSI, and not in a collective determination by PSI's membership or its leadership group.

In connection with principles 2 and 3, it would be desirable to use international agreements to justify or support PSI actions, including interdiction. For example, in 2002, the United States initiated a three-year process at the International Maritime Organization (IMO) to negotiate and develop a more effective international framework for combating maritime terrorism. In October 2005, new protocols to the 1988 Convention for the Suppression of Unlawful Acts of Violence Against the Safety of Maritime Navigation (called "the SUA Convention") were adopted, although these 2005 protocols are yet to enter into force as of May 2008. The new protocols state that a party commits an offence by engaging in various specific acts, the most pertinent of which to PSI is one involving "transports on board a ship [of] any equipment, materials or software or related technology that significantly contributes to the design, manufacture or delivery of a BCN [biological, chemical, or nuclear] weapon, with the intention that it will be used for such purpose."[5] The new SUA protocols also contain a provision for ship-boarding stating that, subject to the approval of the ship's flag state, another state party can board and search the ship, its cargo, and persons on board. These protocols offer justification and support for potential PSI interdictions in similar situations.

It is easier for countries to join an international agreement to take actions against non-state actors than against states of proliferation concern. The non-state actors are viewed as terrorists, and countries have little hesitation to fight them. However, for various political and commercial reasons, as we have described,[6] countries may be reluctant to offend countries of "proliferation concern." This observation suggests that a two-prong approach is needed to seek support from international agreements for PSI. First, the international anti-terrorism agreements must be drawn on to support PSI situations that are in effect dealing with non-state actors of "proliferation concern." The SUA example (see paragraph directly above) covers this approach.

Second, it must be pointed out that the anti-terrorism provision may not be applicable to states of "proliferation concern" in some

---

5 International Maritime Organization, "Convention for the Suppression of Unlawful Acts Against the Safety of Maritime Navigation," 1988.

6 See, for example, the discussion of China's relationship with North Korea under "Costs and Benefits of China's PSI Affiliation" in Chapter Two.

instances and that the issue will instead have to be directly dealt with. For the SUA example given above, one should not extrapolate these protocols to situations in which passage is not innocent by arguing that consent from the flag state to search the ship should be required. In the case of SUA, the flag state would be unaware of the illicit transport (a terrorist act) and would appreciate another party's assistance with the inspection. In the PSI case, there needs to be a way to inspect a ship that belongs to a state of proliferation concern or to a state willing to assist a proliferator for commercial or other reasons. The best place to justify PSI interaction with a ship is in the waters of a PSI-affiliated country. The most direct way to do this is to make the interpretation of the right of innocent passage unambiguous in the UNCLOS.[7]

In sum, we encourage the use of international agreements to justify PSI activities or to carry out actions that are parallel to PSI purposes.[8]

Principle 4

The idea behind principle 4 is that PSI affiliation is only a modest increment to the five countries' earlier undertakings and complements prior undertakings directed at preventing proliferation. For example, each country has already reported on implementation measures it has taken

---

[7] President Clinton signed the treaty in 1994, and the Bush administration supports the treaty and has been pushing for its ratification. There have long been controversies over why the United States should or should not ratify it. As far as the implications of ratification for PSI are concerned, the benefit to PSI would be the largest if the Senate were to ratify it and the United States were actively involved in promoting the interpretation of right of innocent passage, including conditions under which the right should be voided, as we have described in this report.

[8] While the following example is not related to principles 2 and 3, it is broadly supportive of PSI. Such international agreements as the Australia Group have a catch-all control similar to one in the U.S. domestic export control framework. The countries participating in these international agreements want exporters to notify the authorities if the exporters are aware that the nonlisted items may contribute to proscribed activities. This catch-all provision allows the international community to cast a much wider net for transfers of illicit WMD items. This provision can reduce the illicit WMD traffic and make PSI's task relatively easier, because it makes it more difficult for a proliferator to acquire a compliant exporter. Moreover, once such undeclared items are discovered during a PSI interdiction, the offending exporter would have fewer excuses and would more likely be punished.

or intends to take in response to UNSCR 1540's call to prohibit non-state actors from developing, acquiring, or transporting WMD/missile items. These reports show that these five countries have already made significant efforts to enact domestic laws and join international treaties and agreements for nonproliferation. Affiliation with PSI can thus be viewed as a modest increment and complement to already endorsed nonproliferation efforts and objectives. This view will gain in credibility and emphasis if it is jointly conveyed to each of the five countries by several OEG-member countries.

Principle 5

The idea behind principle 5 is that the stature of China, India, Indonesia, and Pakistan in the global arena warrants an invitation to affiliate with PSI that includes the option of immediate membership in the OEG. As members in the OEG, which oversees PSI's ongoing agenda and guides its activities, including exercises, these countries would be able to influence how PSI is applied, as well as represent their own interests and concerns.

Applying the Principles

Indonesia and Malaysia

To enhance the prospects for PSI affiliation, it would be advantageous not to have the initial invitation to Indonesia and Malaysia—or the resumption of approaches that may have already occurred—be undertaken and followed up by the United States. Instead, it might be preferable for the invitation process to be led by three or four PSI affiliates that have good and close relationships with Indonesia and Malaysia, relationships not marred or complicated by U.S. dominance. For example, the PSI invitations might come from Singapore, Japan, France, Australia, and one or two states in the Gulf Cooperation Council (GCC) (principle 1). Acting on behalf of the full PSI constituency, these countries would explicate the subjectivity of determining what may or may not be innocent passage (principle 2) and the unambiguous protection of territorial waters by the littoral states (principle 3). New Zealand might usefully be included among the several soliciting countries, partly because of its geographic proximity and partly because it has effectively articulated the broad scope of benefits from affiliation.[9]

Indonesia and Malaysia should be assured that, according to PSI's fundamental tenets, no PSI affiliate can interdict ships in their territorial waters without their approval. Hence, Indonesia and Malaysia need not be concerned that PSI affiliation will entail any infringement of sovereignty. Nor would PSI affiliation oblige them to interdict any foreign ship in their waters or elsewhere if they wish not to. Finally, as affiliates of PSI, their flagged ships will not be interdicted without their consent.

For reasons mentioned in Chapter One, this approach may be conducive to Indonesia and Malaysia's reconsideration of their prior estimates of the costs and benefits of PSI affiliation. Moreover, this reconsideration is less likely to be hindered by excessive concern that sovereignty or national sensitivity might be compromised by affiliation. Also, for reasons mentioned earlier, it may be advisable to pursue this approach with Malaysia first, followed by Indonesia. Despite the United States' limited and less conspicuous role in the invitation process, proceeding in this sequential manner may avoid any residual anti-American sensitivity that might intrude in Indonesia but would be attenuated were Malaysia inclined to proceed in a favorable direction. Moreover, approaching Malaysia first takes advantage of Malaysia's already having been an observer in three PSI exercises and having joined CSI.

This case is also an application of principle 4 for building on relevant prior activities by the five countries, including their recent efforts in enacting domestic laws and joining international agreements for

---

[9] New Zealand's minister of foreign affairs made the following statement: "New Zealand believes that terrorism and 'rogue states' are among the greatest current threats to peace and stability. . . . It [PSI] has forged a strong network of cooperation among members. This enables effective information-sharing, not only between the governments and security agencies of different member countries, but also between governments and the private sector, for example, the transport industry." Winston Peters (Minister of Foreign Affairs), New Zealand Ministry of Foreign Affairs and Trade, "PSI Initiative Tackles Trade in Weapons of Terror," March 26, 2007.

nonproliferation. While Indonesia has not observed any PSI exercises or joined CSI, it has recently made similar efforts domestically and internationally in support of nonproliferation. Moreover, because of Indonesia's stature and strategic location, its invitation should include the option of immediate membership in the OEG.

India and Pakistan

It may be advisable to have the three non-U.S. nuclear states that are affiliated with PSI—France, the United Kingdom, and Russia—and perhaps Japan as well, convey invitations to India and Pakistan. With the United States not in the role of formal protagonist (principle 1), India's sensitivity (and, more specifically, the acute anti-U.S. hostility associated with the Indian Communist Party, and India's contemporaneous consideration of the NPT with the United States) may be allayed. Similarly, the prospects for India's affiliation would be enhanced if the invitation follow-up is managed by these prominent powers while the United States plays a less conspicuous, supporting role.

For India and Pakistan (and the other three countries, as well), the protagonists should assure that PSI will not compromise their right of innocent passage (principle 2) and should emphasize the incremental character of PSI affiliation in light of these countries' prior efforts to enact domestic laws and to join international agreements for nonproliferation (principle 4)—for example, the fact that India and Pakistan have already participated as observers in PSI exercises. Finally, it would be both appropriate and perhaps effective to extend with the invitation to both India and Pakistan the option of immediate membership in the OEG (principle 5).

China

The connections, interactions, and interests that both link China with and distance China from PSI have already been described. Given that China is a nuclear power with serious interests in nonproliferation and in countering terrorism and terrorists' access to nuclear components, the invitation for affiliation would best be handled by several principal PSI members, including but not confined to the United States. Further, the invitation's effectiveness would be enhanced if the United States played a lesser role than the others (principle 1). For example, France, the United Kingdom, Germany, and the United States could jointly offer the invitation, but the first three countries would play the lead role. France, and perhaps Russia as another PSI member, might authoritatively convey the consistency that exists between, on one hand, PSI affiliation and, on the other, the appropriate and reasonable qualifications within PSI's interdiction principles that can be invoked to protect the right of genuinely "innocent" passage (principle 2).

In the case of China, the United States may be in the best position to explicate the incremental and complementary nature of PSI affiliation (principle 4). China has joined many international nonproliferation treaties and agreements in the last two decades, and it has already placed three of its principal ports (Hong Kong, Shanghai, and Shenzhen) under CSI. Consequently, PSI affiliation can be presented as a modest additional step that complements China's other nonproliferation efforts. The invitation's persuasiveness is likely to be enhanced by having all four of the inviting powers extend to China the option of immediate status as part of the OEG upon affiliation (principle 5).

CHAPTER FOUR

Preliminary Ideas for Further Consideration

We have begun thinking about further ideas that might be developed and steps that might be taken to enhance PSI's inclusiveness and effectiveness in the future. We summarize here a few preliminary ideas that may warrant further consideration.

- *Discuss with the insurance industry whether and, if so, how premiums charged for insuring cargo (whether transported by surface, air, or sea) take into account affiliation with PSI of the transport vehicle's nation of origin.* The idea underlying this inquiry is that PSI participation may reduce various risks, such as those associated with possible transport of WMD/missile items and with possible interdiction delays relating to such cargo.

The insurance problem is complicated. On one hand, to the extent that the 91 participants in PSI create a form of deterrence against WMD proliferation, this "public good" benefits all transport, whether the shipper's flag is that of a PSI affiliate or not. This line of reasoning might suggest that insurance premiums for flagged vehicles should not differ on the basis of PSI affiliation. On the other hand, however, some benefits of PSI accrue more directly to affiliates. As a result of the affiliated countries' improved customs procedures, inspection capabilities, and shared intelligence, their flagged vehicles are more likely to be "clean" and "safe" than are those of unaffiliated countries. Anticipating these circumstances, potential state or non-state proliferators may be more likely to choose the flagged vehicles of the unaffiliated. Consequently, it may be inferred that the risk exposure of the unaffiliated—in the form, for example, of accidents in transit or of interdiction or interruption and hence delays in transit—may be calculably greater than that of PSI affiliates.

In effect, the transport of WMD/missile items involves increased risks to the points of origin, destination, and thoroughfare. In turn, PSI affiliation carries with it some presumptive reduction of these risks. Do insurance underwriters consider this in deciding whether or not to insure, and what to charge?

More generally, we suggest it would be worthwhile to analyze how these relevant and ostensibly reduced risks might be quantified in furtherance of PSI's broad objectives.

• *Consider ways to allay concerns about the right of innocent passage.* Part of the concern about the right of innocent passage—a concern common to all five countries (as well as others)—is that even an innocent ship might suffer delay because of interdiction. One possible approach would be for PSI affiliates to establish an "interdiction compensation fund" that would be used to compensate the owner of a carrier (ship, ground vehicle, or airplane) and the affected cargo recipients in the event that an interdiction turned out to be based on erroneous intelligence and through no fault of the carrier. The intent of the compensation would be to cover basic losses resulting from the delay in cargo delivery, thereby precluding any windfalls resulting from such interdictions.

It would be useful to consider whether this type of fund would serve the purpose, how it might be structured and financed, and how it would operate. One also needs to consider how legal authorities can be established so that the potential compensation would preclude frivolous lawsuits and claims. Consideration of these issues should be pursued multilaterally within PSI, perhaps under OEG auspices.

• *Clarify possible misinterpretation about the relationship between UNCLOS and PSI with respect to the right of innocent passage, including appropriate rules of engagement that would reassure littoral states that their prerogatives in their own territorial seas would not be infringed upon by PSI interdiction principles.* The PSI interdiction principles (see Appendix B) include the following paragraphs:

4. To take specific actions in support of interdiction efforts regarding cargoes of WMD, their delivery systems, or related materials, to the extent their national legal authorities permit and consistent with their obligations under international law and frameworks, to include:

d. To take appropriate actions to (1) stop and/or search in their internal waters, territorial seas, or contiguous zones (when declared) vessels that are reasonably suspected of carrying such cargoes to or from states or non-state actors of proliferation concern and to seize such cargoes that are identified; and (2) to enforce conditions on vessels entering or leaving their ports, internal waters or territorial seas that are reasonably suspected of carrying such cargoes, such as requiring that such vessels be subject to boarding, search, and seizure of such cargoes prior to entry.

These paragraphs imply that a PSI-affiliated country may interdict a foreign flagged ship within its territorial seas that is reasonably suspected of carrying WMD/missile items, provided the interdiction is consistent with the country's "obligations under international law and frameworks." In consequence, a PSI-affiliated nation, B, may have a plausible concern that a PSI-affiliated littoral state, A, might interpret these obligations to mean it can interdict B's flag ship without B's consent so long as the interdiction is conducted in A's territorial waters and is preceded by reasonable grounds for suspicion. In practice, however, the flag ship of B would not be interdicted by A—either in A's territorial seas or on the high seas—without B's permission. PSI's OEG should make this explicit, thereby helping to resolve what several nations that might otherwise affiliate with PSI see as a key obstacle.

Further, PSI might encourage B to offer an alternative to outright refusal to have its suspected ship inspected by A. For example, B might offer to have its suspected ship inspected outside A's territorial seas, in the presence of an observer from A or a third-party PSI member chosen by B. Since B's consent would still be necessary, B might be convinced that its ship would have a better chance of being satisfactorily inspected.

• *Consider whether to offer prospective PSI participants technical assistance, inspection equipment, and other items that might help them improve their import/export control, inspection, and interdiction capabilities.* We do not think the five countries should be offered appreciable amounts of assistance and equipment—in our view, the primary inducement for affiliating with PSI should not be indirect benefits but, instead, benefits that are directly connected with affiliation. Nevertheless, low-cost forms of assistance that directly contribute to PSI objectives might be considered as a further inducement for affiliation with PSI.

• *Analyze the status and trends of technology for detecting and sensing WMD that may enable better and quicker identification of WMD components.* Improvements in this technology may enable more-accurate and more-rapid identification of WMD components, thereby enhancing the future effectiveness of PSI.

APPENDIX A

The Proliferation Security Initiative (PSI)

The following is a (reformatted) replication of the Initiative's text provided by the U.S. Department of State at http://www.state.gov/t/isn/rls/fs/105217.htm

Fact Sheet
Bureau of International Security and Nonproliferation
Washington, DC
May 26, 2008

The Proliferation Security Initiative (PSI)

Previous Version: 2006

What Is the Proliferation Security Initiative?

The Proliferation Security Initiative (PSI) is a global effort that aims to stop trafficking of weapons of mass destruction (WMD), their delivery systems, and related materials to and from states and non-state actors of proliferation concern. Launched by President Bush on May 31, 2003, U.S. involvement in the PSI stems from the U.S. National Strategy to Combat Weapons of Mass Destruction issued in December 2002. That strategy recognizes the need for more robust tools to stop proliferation of WMD around the world, and specifically identifies interdiction as an area where greater focus will be placed. Today, more than 90 countries around the world support the PSI.

The PSI is an innovative and proactive approach to preventing proliferation that relies on voluntary actions by states that are consistent with national legal authorities and relevant international law and frameworks. PSI participants use existing authorities—national and international—to put an end to WMD-related trafficking and take steps to strengthen those authorities as necessary. UN Security Council Resolution 1540, adopted unanimously by the Security Council, called on all states to take cooperative action to prevent trafficking in WMD. The PSI is a positive way to take such cooperative action.

In September 2003, PSI participants agreed to the PSI Statement of Interdiction Principles that identifies specific steps participants can take to effectively interdict WMD-related shipments and prevent proliferation. The PSI Principles also recognize the value in cooperative action and encourage participating countries to work together to apply intelligence, diplomatic, law enforcement, military, and other capabilities to prevent WMD-related transfers to states and non-state actors of proliferation concern.

PSI partners encourage all states to endorse the PSI, and to take the steps outlined in the Principles. Support for the PSI is an acknowledgment of the need for stronger measures to defeat proliferators through cooperation with other countries.

What is the value of the PSI?

The PSI provides committed states with a framework for coordinating counterproliferation activities to thwart proliferators' increasingly sophisticated tactics. In recent years, we have seen the emergence of black-market operatives who, for the right price, are willing to use their knowledge, access to materials, and personal connections to provide WMD-related goods and services to terrorists and countries of proliferation concern. Five years ago, the world became aware that an international black market network, headed by Dr. A.Q. Khan, had for many years been supplying clandestine nuclear weapons programs. Seizure of the cargo ship BBC China exposed the network and ultimately led to Libya's decision to end its nuclear and missile programs. Most recently, the discovery of Syria's covert nuclear reactor—believed not to be for peaceful purposes—demonstrated that proliferators are capable of pursuing their dangerous objectives even as the world is watching. And today, Iran continues its pursuit of nuclear technology and missile systems that could deliver WMD in direct violation of the UN Security Council.

Proliferators and their facilitators continue to work aggressively to circumvent export controls, establish front companies to deceive legitimate firms into selling them WMD-related materials, ship WMD-related materials under false or incomplete manifests, and launder their financial transactions through established banking institutions. These proliferation activities undermine international peace and security and require an international response.

While states have cooperated for many years to combat WMD proliferation and prevent specific shipments of WMD, their delivery systems, and related materials, these efforts have largely been ad hoc. The PSI takes these efforts out of the ad hoc realm by facilitating information-sharing, building relationships between international counterparts at the political and operational levels, and providing a forum for experts to share best practices on organizing for and conducting interdictions.

Our deeper understanding of today's proliferation threat has increased international support, including widespread attention at senior levels of government, for more concerted efforts to halt WMD trafficking at all points along the proliferation supply chain. The PSI builds on our interdiction experience to date and uses the full range of counterproliferation tools—diplomacy, intelligence, customs authorities, law enforcement, military, and financial—to meet this pressing challenge.

How Does the PSI Work?

The PSI works in three primary ways. First, it channels international commitment to stopping WMD-related proliferation by focusing on interdiction as a key component of a global counterproliferation strategy. Endorsing the PSI Statement of Interdiction Principles provides a common view of the proliferation problem and a shared vision for addressing it.

Second, the PSI provides participating countries with opportunities to improve national capabilities and authorities to conduct interdictions. A robust PSI exercise program allows participants increase their interoperability, improve interdiction decision-making processes, and enhance the interdiction capacities and readiness of all participating states. In five years, PSI partners have sustained one of the only global, interagency, and multinational exercise programs, conducting over 30 operational air, maritime, and ground interdiction exercises involving over 70 nations. These exercises are hosted throughout the world by individual PSI participants and consist of air, maritime, and ground exercises executed by participants' interagency and ministries focusing on improving coordination mechanisms to support interdiction-related decision-making.

Furthermore, the PSI Operational Experts Group (OEG), a group of military, law enforcement, intelligence, legal, and diplomatic experts from twenty PSI participating states, meets regularly to develop operational concepts, organize the interdiction exercise program, share information about national legal authorities, and pursue cooperation with key industry sectors. The OEG works on behalf of all PSI partners and works enthusiastically to share its insights and experiences through bilateral and multilateral outreach efforts.

Third, and of the most immediate importance, the PSI provides a basis for cooperation among partners on specific actions when the need arises. Interdictions are information-driven and may involve one or several participating states, as geography and circumstances require. The PSI is not a formal treaty-based organization, so it does not obligate participating states to take specific actions at certain times. By working together, PSI partners combine their capabilities to deter and stop proliferation wherever and whenever it takes place.

How Can States Participate in the PSI?

States can become involved in the PSI in multiple ways.

- Formally committing to and publicly endorsing the PSI and the Statement of Interdiction Principles, and indicating willingness to take all steps available to support PSI efforts.
- Undertaking a review and providing information on current national legal authorities to undertake interdictions at sea, in the

air, or on land, and indicating willingness to strengthen authorities, where appropriate.

- Identifying specific national "assets" that might contribute to PSI efforts (e.g., information sharing, military, and/or law enforcement assets).
- Providing points of contact for PSI assistance requests and other operational activities, and establishing appropriate internal government processes to coordinate PSI response efforts.
- Being willing to actively participate in PSI interdiction training exercises and actual operations as opportunities arise.
- Being willing to conclude relevant agreements (e.g., boarding arrangements) or otherwise to establish a concrete basis for cooperation with PSI efforts.

Cooperation by flag, coastal, or transshipment states, and states along major air shipment corridors is particularly essential to counterproliferation efforts involving cargoes in transit.

What Is the Future of the PSI?

The PSI is an enduring initiative that continues to establish a web of counterproliferation partnerships to prevent trade in WMD, their delivery systems and related materials.

By cooperating through PSI, states make it more difficult and costly for proliferators to engage in this deadly trade. Over time, proliferators and others involved in supporting proliferation activities will learn that there are countries determined to work together to take all possible steps to stop their efforts. PSI is an important contribution to global nonproliferation efforts and is a strong deterrent to proliferation-related trafficking. PSI also seeks enhanced export control, regulatory systems, and law enforcement cooperation to shut down proliferation-related networks and activities to bring down those involved to justice.

The United States will work to maintain and build on past PSI successes, including through further development of real-world partnerships, networks of expert contacts, and operational readiness to conduct cooperative interdictions of WMD-related shipments. We will seek to further develop international law enforcement cooperation and will increase our dialogue and cooperation with industry. The United States will also continue to cooperate with our PSI partners to put in place smooth, effective communications and operational procedures.

Rogue states, terrorist and criminal organization, and unscrupulous individuals who contemplate trafficking in WMD related materials must now contend with an international community united in detecting and interdicting such transfers by air, land, and sea.

The PSI participating states encourage endorsement of the Statement of Interdiction Principles and participation in the PSI by all states that are committed to preventing the proliferation of WMD, their means of delivery, and related materials.

For more information on the PSI, see http://www.state.gov/t/isn/c10390.htm.

APPENDIX B

Interdiction Principles of the Proliferation Security Initiative

The following is a (reformatted) replication of the PSI statement of interdiction principles provided by the U.S. Department of State at http://www.state.gov/t/isn/rls/fs/23764.htm

FACT SHEET
The White House, Office of the Press Secretary
Washington, DC
September 4, 2003

Proliferation Security Initiative: Statement of Interdiction Principles

The Proliferation Security Initiative (PSI) is a response to the growing challenge posed by the proliferation of weapons of mass destruction (WMD), their delivery systems, and related materials worldwide. The PSI builds on efforts by the international community to prevent proliferation of such items, including existing treaties and regimes. It is consistent with and a step in the implementation of the UN Security Council Presidential Statement of January 1992, which states that the proliferation of all WMD constitutes a threat to international peace and security, and underlines the need for member states of the UN to prevent proliferation. The PSI is also consistent with recent statements of the G8 and the European Union, establishing that more coherent and concerted efforts are needed to prevent the proliferation of WMD, their delivery systems, and related materials. PSI participants are deeply concerned about this threat and of the danger that these items could fall into the hands of terrorists, and are committed to working together to stop the flow of these items to and from states and non-state actors of proliferation concern.

The PSI seeks to involve in some capacity all states that have a stake in nonproliferation and the ability and willingness to take steps to stop the flow of such items at sea, in the air, or on land. The PSI also seeks cooperation from any state whose vessels, flags, ports, territorial waters, airspace, or land might be used for proliferation purposes by states and non-state actors of proliferation concern. The increasingly aggressive efforts by proliferators to stand outside or to circumvent existing nonproliferation norms, and to profit from such trade, requires new and stronger actions by the international community. We look forward to working with all concerned states on measures they are able and willing to take in support of the PSI, as outlined in the following set of "Interdiction Principles."

Interdiction Principles for the Proliferation Security Initiative

PSI participants are committed to the following interdiction principles to establish a more coordinated and effective basis through which to impede and stop shipments of WMD, delivery systems, and related materials flowing to and from states and non-state actors of proliferation concern, consistent with national legal authorities and relevant international law and frameworks, including the UN Security Council. They call on all states concerned with this threat to international peace and security to join in similarly committing to:

1. Undertake effective measures, either alone or in concert with other states, for interdicting the transfer or transport of WMD, their delivery systems, and related materials to and from states and non-state actors of proliferation concern. "States or non-state actors of proliferation concern" generally refers to those countries or entities that the PSI participants involved establish should be subject to interdiction activities because they are engaged in proliferation through: (1) efforts to develop or acquire chemical, biological, or nuclear weapons and associated delivery

systems; or (2) transfers (either selling, receiving, or facilitating) of WMD, their delivery systems, or related materials.

2. Adopt streamlined procedures for rapid exchange of relevant information concerning suspected proliferation activity, protecting the confidential character of classified information provided by other states as part of this initiative, dedicate appropriate resources and efforts to interdiction operations and capabilities, and maximize coordination among participants in interdiction efforts.
3. Review and work to strengthen their relevant national legal authorities where necessary to accomplish these objectives, and work to strengthen when necessary relevant international law and frameworks in appropriate ways to support these commitments.
4. Take specific actions in support of interdiction efforts regarding cargoes of WMD, their delivery systems, or related materials, to the extent their national legal authorities permit and consistent with their obligations under international law and frameworks, to include:
   a. Not to transport or assist in the transport of any such cargoes to or from states or non-state actors of proliferation concern, and not to allow any persons subject to their jurisdiction to do so.
   b. At their own initiative, or at the request and good cause shown by another state, to take action to board and search any vessel flying their flag in their internal waters or territorial seas, or areas beyond the territorial seas of any other state, that is reasonably suspected of transporting such cargoes to or from states or non-state actors of proliferation concern, and to seize such cargoes that are identified.
   c. To seriously consider providing consent under the appropriate circumstances to the boarding and searching of its own flagged vessels by other states, and to the seizure of such WMD-related cargoes in such vessels that may be identified by such states.

d. To take appropriate actions to (1) stop and/or search in their internal waters, territorial seas, or contiguous zones (when declared) vessels that are reasonably suspected of carrying such cargoes to or from states or non-state actors of proliferation concern and to seize such cargoes that are identified; and (2) to enforce conditions on vessels entering or leaving their ports, internal waters or territorial seas that are reasonably suspected of carrying such cargoes, such as requiring that such vessels be subject to boarding, search, and seizure of such cargoes prior to entry.
e. At their own initiative or upon the request and good cause shown by another state, to (a) require aircraft that are reasonably suspected of carrying such cargoes to or from states or non-state actors of proliferation concern and that are transiting their airspace to land for inspection and seize any such cargoes that are identified; and/or (b) deny aircraft reasonably suspected of carrying such cargoes transit rights through their airspace in advance of such flights.
f. If their ports, airfields, or other facilities are used as transshipment points for shipment of such cargoes to or from states or non-state actors of proliferation concern, to inspect vessels, aircraft, or other modes of transport reasonably suspected of carrying such cargoes, and to seize such cargoes that are identified.

APPENDIX C

Status of Countries on Treaties, Agreements, and Programs of Potential Relevance to PSI

We present here, in tabular form, the current status of the five countries of interest—Malaysia, Indonesia, Pakistan, India, and China—and the United States with respect to various treaties, agreements, and programs that are potentially relevant to PSI and thus to the countries' potential for affiliating with PSI. To provide a quick, overall picture of an individual country's commitment to nonproliferation, the forms of participation and nonparticipation noted in the columns are differentiated: Those that are unshaded indicate that the country is generally acting in accord with the intent of a treaty, agreement, or program; those that are shaded indicate nonparticipation.

Status of Countries on Treaties, Agreements, and Programs Potentially Relevant to PSI

	Five Countries					
Treaty, Agreement, or Program	**Indonesia**	**Malaysia**	**Pakistan**	**India**	**China**	**U.S.**
PSI affiliation	No	No	No	No	No	M
PSI exercise	No	Yes	Yes	Yes	No	Yes
Nuclear Suppliers Group	No	No	No	P	M	M
Zangger Committee	No	No	No	No	M	M
Wassenaar Arrangement	No	No	No	No	No	M
Australia Group	No	No	No	No	No	M
Missile Technology Control Regime	No	No	No	No	Ab	M
The Hague Code of Conduct Against Ballistic Missiles	No	No	No	No	No	Si
UNSCR 1540	RS	RS	RS	RS	RS	RS
Nuclear Nonproliferation Treaty	SP	SP	No	No	SP	SP
Comprehensive Nuclear Test Ban Treaty	SP	St	No	No	St	St
Partial Test Ban Treaty	SP	SP	SP	SP	No	SP
IAEA Safeguard Agreement	Yes	Yes	Yes	Yes	Yes	Yes
Convention on the Physical Protection of Nuclear Material	SP	Ac	SP	Ac	SP	SP
Chemical Weapon Convention	SP	SP	SP	SP	SP	SP
Biological Weapon Convention	SP	SP	SP	SP	SP	SP

Status of Countries on Treaties, Agreements, and Programs Potentially Relevant to PSI—continued

Treaty, Agreement, or Program	Five Countries					
	Indonesia	Malaysia	Pakistan	India	China	U.S.
Biological and Toxic Weapons Convention Confidence-Building Measures	No	NSD	SSD	SSD	SSD	SSD
Suppression of Acts of Nuclear Terrorism	No	No	No	SP	SP	Si
Container Security Initiative	No	Yes	No	No	Yes	Yes
Export Control and Related Border Security Assistance Program	Yes	Yes	Yes	Yes	Yes	Yes
UNCLOS 1982	SP	SP	SP	SP	SP	St
Suppression of Acts of Nuclear Terrorism	No	No	No	SP	SP	Si

NOTE: Unshaded entries in the columns indicate that the country's actions are in general accord with the intent of the treaty, agreement, or program; shaded entries indicate nonparticipation. Entries are defined as follows: Ab = abides by but has not signed or ratified; Ac = has acceded (has consented to become a party but has not signed); M = member; No = has not acceded, signed, or ratified; NSD = has submitted no annual data; P = pending; RS = has submitted report; Si = has signed; SP = state party (signed and ratified); SSD = has submitted at least some annual data; St = signatory (signed but not ratified); Yes = has participated.

APPENDIX D

Summary of Pakistan's Report on National Measures on the Implementation of Security Council Resolution 1540 (2004)

Pakistan's report on the national measures it has taken in implementing UNSCR 1540 was submitted on October 27, 2004, to the chairman of the Security Council committee established pursuant to resolution 1540 (2004). Our summary of the report is as follows:

- Pakistan declares full support of effective measures to prevent non-state actors from accessing WMD/missile items.
- Pakistan stands ready to strengthen global nonproliferation within existing treaty regimes and bodies.
- Pakistan states that it has legal, administrative, and law enforcement measures in place to prohibit non-state actors' access to WMD/missile items.
- Pakistan has legal and administrative instruments for export control.
- Pakistan passed a new comprehensive national legislation, Export Control on Goods, Technologies, Material and Equipment Related to Nuclear and Biological Weapons and Their Delivery systems Act, 2004. This occurred soon after the A. Q. Khan nuclear smuggling was discovered. Moreover, the United States and Japan assisted in drafting the legislation. This also indicates that export and import laws and inspections, and domestic control of WMD/missile items are areas in which the United States

can offer more assistance for inducing Pakistan (and other countries) to join PSI.
- Pakistan has laws to effectively account for and secure sensitive materials in production, use, storage, or transport.
- Pakistan has an established system for the safety and security of nuclear and radioactive materials.
- Pakistan's customs and other law enforcement agencies keep a close watch at borders, seaports, and airports.
- Special training to detect sensitive materials is being provided to customers and law enforcement officials.
- Pakistan states that it is in a position to assist other countries in implementing UNSCR 1540 within their territories, such as establishing a legal and regulatory infrastructure for controlling WMD/missile items.

References

"Agreement for Cooperation Between the Government of the United States of America and the Government of India Concerning Peaceful Uses of Nuclear Energy," August 3, 2007.

"China, Iran Sign Biggest Oil and Gas Deal," *China Daily*, October 31, 2004. As of September 25, 2008:
http://www.chinadaily.com.cn/english/doc/2004-10/31/content_387140.htm

"CPI(M)'s Open Letter to Members of Parliament," *People's Democracy* (Weekly Organ of the Communist Party of India (Marxist), Vol. XXXI, No. 38, September 23, 2007.

Gordon, Michael R., "Indonesian Scolds U.S. on Terrorism Fight," *New York Times*, June 7, 2006.

Hachigian, Nina, and Mona Sutphen, *The Next American Century*, 2008, pp. 142–145.

Hawkins, William, "Chinese Realpolitik and the Proliferation Security Initiative," Association for Asia Research, February 18, 2005.

[The] Henry J. Hyde United States–India Peaceful Atomic Energy Cooperation Act of 2006, Public Law 109-401, 109th Congress, 1st Sess., December 18, 2006. As of September 25, 2008:
http://frwebgate.access.gpo.gov/cgi-bin/getdoc.cgi?dbname=109_cong_public_laws&docid=f:publ401.109

Ibrahim, Youssef, "Saudis Warn They May Use Missiles Against Iran," *New York Times*, April 29, 1988.

"India Says No to Logistics Deal with US, for Now," *Times of India*, February 27, 2008.

"Indian Government Boosts Maritime Security; Stays Cool to Proliferation Security Initiative," *Asian Export Control Observer*, 6, February/March 2005, p. 9.

"Indonesia Questions US Proposals on Proliferation Security Initiative," *Jakarta Post*, March 16, 2006. As of September 25, 2008:
http://www.redorbit.com/news/international/430553/indonesia_questions_us_proposals_on_proliferation_security_initiative/index.html).

International Maritime Organization, "Convention for the Suppression of Unlawful Acts Against the Safety of Maritime Navigation," 1988. As of September 25, 2008:
http://www.imo.org/Conventions/mainframe.asp?topic_id=259&doc_id=686

"Iran, Malaysia Sign $16B Oil Deal," *China Daily*, December 27, 2007. As of September 25, 2008:
http://www.chinadaily.com.cn/world/2007-12/27/content_6351901.htm

"Israel: Saudi Arabia's Purchase of Nuclear Warheads Said 'Threat to World Peace,'" translated excerpt, *Foreign Broadcast Information Service*, October 22, 2003.

James Martin Center for Nonproliferation Studies, "NAM Chair Malaysia Skeptical of UNSC Involvement," January 2006. As of September 25, 2008:
http://cns.miis.edu/research/iran/reaction/malaysia.htm

Lieggi, Stephanie, "Proliferation Security Initiative Exercise Hosted by Japan Shows Growing Interest in Asia but No Sea Change in Key Outsider States," *WMD Insights*, December 2007–January 2008. As of October 22, 2008, issue available in PDF through:
http://www.wmdinsights.com/archive.htm

"Malaysia Arrests Alleged Black Market Nuclear Agent," *Taipei Times*, May 30, 2004.

"Malaysia in No Hurry to Join US-led Security Pact," *Reuters*, April 17, 2007. As of September 25, 2008:
http://www.reuters.com/article/topNews/idUSKLR21761620070417?feedType=RSS

"Malaysia Still Studying Membership in PSI, Says Najib," *Malaysian National News Agency*, BERNAMA.com, April 17, 2007.

Ministry of Foreign Affairs of Japan, *PSI Maritime Interdiction Exercise "Pacific Shield 07" Hosted by the Government of Japan (Overview and Evaluation)*, October 18, 2007. As of September 25, 2008:
http://www.mofa.go.jp/policy/UN/disarmament/arms/psi/overview0710.html

Ministry of Foreign Affairs of the People's Republic of China, "Foreign Ministry Spokesperson's Press Conference on 4 December 2003," December 4, 2003. As of September 23, 2008:
http://www.fmprc.gov.cn/eng/xwfw/2510/2511/t55556.htm

———, "Foreign Ministry Spokesperson Zhang Qiyue's Press Conference on 26 October 2004," October 27, 2004. As of September 23, 2008:
http://www.fmprc.gov.cn/eng/xwfw/s2510/2511/t167984.htm

———, "Proliferation Security Initiative," May 21, 2007. As of September 23, 2008:
http://www.fmprc.gov.cn/eng/wjb/zzjg/jks/kjlc/fkswt/dbfks/t321019.htm

National Defense Authorization Act for Fiscal Year 2000, Public Law 106-65, 106th Congress, October 5, 1999.

Nojima, Tsuyoshi, "Japan to Offer Patrol Ships to Curb Piracy," *Asahi Shimbun*, March 17, 2005 (in Lexis-Nexis Academic Universe: http://www.lexis-nexis.com), as cited in "Chinese Port of Shanghai Joins U.S. Container Security Initiative; Argentina and Brazil to Follow Suit," *Asian Export Control Observer*, 7, April/May 2005, p. 8. As of October 22, 2008, issue available in PDF through:
http://cns.miis.edu/pubs/observer/asian/

"Pakistan Joins Initiative to Combat Nuclear Terrorism, Establishes Strategic Export Control Division," *International Export Control Observer*, 11, June/July 2007, p. 2. As of September 25, 2008:
http://cns.miis.edu/pubs/observer/

Pakistan's National Report on National Measures on the Implementation of Security Council Resolution 1540 (2004), United Nations S/AC.44/2004/(02)/22, October 27, 2004. As of October 22, 2008, available in PDF through:
http://www.un.org/Docs/journal/asp/ws.asp?m=S/AC.44/2004/(02)/22

Peters, Winston (Minister of Foreign Affairs), "PSI Initiative Tackles Trade in Weapons of Terror," New Zealand Ministry of Foreign Affairs and Trade, March 26, 2007. As of September 25, 2008:
http://www.mfat.govt.nz/Foreign-Relations/1-Global-Issues/Security/0-PSI-Meeting-Peters.php

Public Law 106-65—See *National Defense Authorization Act for Fiscal Year 2000*

Public Law 109-401—See *[The] Henry J. Hyde United States–India Peaceful Atomic Energy Cooperation Act of 2006*

Rabasa, Angel, and Peter Chalk, *Indonesia's Transformation and the Stability of Southeast Asia*, MG-1344-AF, Santa Monica, Calif.: RAND Corporation, 2001. As of October 13, 2008:
http://www.rand.org/pubs/monograph_reports/MR1344/

"Security, Peace and Prosperity for All," *Reuters*, February 25, 2008. As of September 25, 2008:
http://www.reuters.com/article/pressRelease/idUS140532+25-Feb-2008+PRN20080225

Sudarsono, Juwono (Indonesia's Defense Minister), "U.S. Secretary of Defense," June 13, 2006. As of September 25, 2008:
http://juwonosudarsono.com/wordpress/?p=4

Suryanarayana, P. S., "The Country Is Enhancing Its Defence Preparedness, Says Minister," *The Hindu*, June 4, 2006.

United Nations Security Council Resolution 1540 (2004), S/RES/1540 (2004), April 28, 2004. As of September 25, 2008, available in PDF through:
http://www.un.org/sc/1540/

"United States Steps Up Pressure on Iran WMD Programs Through Sanctions," *International Export Control Observer*, 4, February 2006, pp. 12–14.

U.S. Department of State, Bureau of International Security and Nonproliferation, *Proliferation Security Initiative: Statement of Interdiction Principles*, Fact Sheet, The White House, Office of the Press Secretary, September 4, 2003. As of September 25, 2008:
http://www.state.gov/t/isn/rls/fs/23764.htm

———, *The Proliferation Security Initiative (PSI)*, Fact Sheet, Bureau of International Security and Nonproliferation,Washington, D.C., May 26, 2008. As of September 25, 2008:
http://www.state.gov/t/isn/rls/fs/105217.htm

Valencia, Mark, "The Proliferation Security Initiative: Making Waves in Asia," *Adelphi Papers*, 45:376, 2005, p. 48.

Weissman, Steve, and Herbert Krosney, *The Islamic Bomb*, Times Books, New York, 1981.